atlas of botany

atlas of botany

ROYAL
BOTANIC
GARDEN
EDINBURGH

8   Foreword

10 Plants and places

ANCIENT BIOMES

16 The first land plants

18 Vascular plants form

20 First complex ecosystems

24 First true forests

28 Coal forests evolve

32 The rise of conifers

36 Ferns and seed plants diversify

40 Plants in the age of dinosaurs

44 Flowering plants emerge

48 Surviving extinction

52 The expansion of grasslands

56 Surviving the Ice Age

MODERN BIOMES

62 Introduction

FORESTS

- 68 Forests
- 70 Adaptations: light and space
- 72 Adaptations: seasons
- 74 Plants of tropical and subtropical moist forests
- 80 Peas
- 82 Heavy metals
- 84 Kapok tree
- 86 Plants of tropical and subtropical dry forests
- 92 Myrtles
- 94 Plants of the cloud forest
- 100 Fighting disease
- 102 Plants of the Mediterranean forest
- 108 Mints
- 110 Plants of the temperate forest
- 116 Bay laurel
- 118 Oaks and allies
- 120 Invasive species
- 122 Plants of the boreal forest
- 128 Twinflower
- 130 Pines
- 132 Cacao

GRASSLANDS AND SHRUBLANDS

- 136 Grasslands and shrublands
- 138 Adaptations: growth and photosynthesis
- 140 Adaptations: fire and predators
- 142 Plants of cold grasslands and shrublands
- 148 Heathers
- 150 Plants of temperate grasslands, savanna, and shrublands
- 156 Yarrow
- 158 Grazing
- 160 Plants of tropical and subtropical grasslands, savanna, and shrublands
- 166 Baobab tree

DESERTS

- 170 Deserts
- 172 Adaptations: form and water
- 174 Adaptations: pollination and defence
- 176 Plants of hot dry deserts
- 182 Cacti
- 184 Prickly pear
- 186 Plants of semi-arid deserts
- 192 Desert greening
- 194 Plants of coastal deserts
- 200 Aloes and allies
- 202 Plants of cold deserts

MOUNTAINS

- 210 Mountains
- 212 Adaptations: altitude and exposure
- 214 Adaptations: isolation and temperature
- 216 Plants of low mountains
- 222 Saxifrages
- 224 Plants of high alpine areas
- 230 Frailejónes
- 232 Rescue missions
- 234 Plants of the páramo
- 240 Primulas

WETLANDS

244 Wetlands
246 Adaptations: air and movement
248 Adaptations: salt and buoyancy
250 Plants of freshwater wetlands
256 Papyrus
258 Pitcher plants
260 Plants of marine and coastal wetlands
266 Giant waterlily
268 Mangroves
270 Algal bloom

HUMAN ENVIRONMENTS

274 Human environments
276 Adaptations: competition and growth
278 Adaptations: crop domestication
280 Plants of urban areas
286 Crops
288 Banana
290 Plants of agricultural areas
296 Imperial crops
298 Pollution
300 Plants of protected areas

306 Glossary
312 Index
318 Acknowledgments

• FOREWORD •

FOREWORD

Plants are astonishing organisms: they define the world as we know it, regulating Earth's water and nutrient cycles, stabilizing the atmosphere, and capturing energy from the Sun to sustain a vast array of life. Every species seems beautifully adapted to its niche, from tiny duckweed in a pond to mighty rainforest kapok trees, and each makes a contribution to the ecosystem. Plants have become adept at making modifications to suit their circumstances – changing the colour or shape of a flower to attract a particular pollinator; harnessing lightning to eliminate competition; adapting their roots or leaves to access alternative sources of nutrition; or providing ants with food and shelter in return for the use of their stings as a defence against herbivory.

This book invites you to explore plant diversity across space and time, following the shifting patterns of habitats around the globe, and celebrates the strategies that enable plants to thrive. The first section delves into the deep past; looking at the rise of plants on land and the traces left as fossils, it reveals how plants shaped early terrestrial landscapes and reinvented themselves in response to extinction events. The second section examines the myriad habitats of our world today, and the often ingenious ways in which plants cope in different environments. It is a grand tour that takes you from Arctic and alpine zones through forests, shrublands, grasslands, and deserts, and into aquatic habitats, as well as to arenas where plants and human interests intersect. Anthropogenic influences affect every biological community, but because humans and plants have shaped each other, knowing more about the "green kingdom" allows us to better understand our place on this shared planet.

Working together
The dynamic relationship between plants and the animal life that shares their space is captured in this depiction of a Cattleya orchid and Brazilian hummingbirds in a detail from a painting by Martin Johnson Heade (1871).

PLANTS AND PLACES

Literally rooted in place, plants have for millions of years confronted the challenge of how to survive in an environment that changes around them. The process of adaptation is never-ending and plants have evolved in many extraordinary ways.

Since they first appeared more than 400 million years ago, land plants have diversified into a vast array of forms, each adapted to cope with the prevailing conditions. Fast-growing lycophytes in the Carboniferous Period, for example, exploited high levels of carbon dioxide in the atmosphere, and locked up vast amounts of carbon inside their woody tissues. They became victims of their own success when this ability lowered atmospheric carbon dioxide, contributing to a global cooling event that brought about their own demise.

PATTERNS OF CHANGE

Across the millennia, countless plant species have arisen, diversified, and adapted to particular conditions, before going extinct when the environment – or other organisms such as their pollinators or herbivores – changed. Today, some of these lost plants are only known from their fossils, but others survive as tantalizing genetic traces in their living relatives.

Biogeography – the study of life in the context of place – reveals how the slow tectonic movement of land masses coupled with chance dispersals of spores or seeds over long distances have created intriguing patterns of species across the globe. These patterns reflect an intricate weaving together of evolution and movements by plants and other organisms over millions of years. The drivers for change have been many and varied, but whether in response to shifting climate conditions, diverse soils, drifting continents, or interactions with each other, every species has evolved into its own niche in the vast tapestry of life.

• INTRODUCTION •

Protea cynaroides

Banksia coccinea

Far-flung family
The Proteas of southern Africa and Banksias of Australia represent two distant relatives that spread through the southern continents, including Antarctica, over 100 million years ago. Despite the separation in their modern distribution, they have developed in ways that are remarkably similar.

Since their transition from water to land, plants have coevolved with myriad other organisms – some as predators, others as partners. They defend themselves against herbivores through vicious spines, shards of glass in their tissues, or a seemingly nefarious array of poisons. However, plants also co-opt pollinators to help them reproduce, often – but not always – rewarding their associates. Almost all land plants have adopted partner organisms to provide nutrients in the form of mycorrhizal fungi that make up the so-called "wood-wide web". In some cases these partnerships are mutually beneficial, in others, the partners are exploited to the plant's advantage.

MOBILE OPERATORS

Although they are unable to run from threats or towards essential nutrients, plants can be surprisingly mobile. Venus flytrap (*Dionaea muscipula*) and sensitive plant (*Mimosa pudica*) are famously fast movers that respond to touch with reflex actions – electrically mediated movements – uncannily like those of animals, while many other plants perform a daily dance as their leaves track the sun, or appear to walk on the shifting sands of beaches. These are just a few of the many ways plants cope with their changing environment.

The greatest plant movements are the dispersals of spores or pollen, fruits, and seeds – perfect packages that can float, glide, fly, or hitch a ride to their destination. The countless dispersal attempts by spores and seeds that allow plants to colonize entirely new areas can, if they succeed, change landscapes and ecosystems forever. The intricate map of plant life on Earth is a tale of these migrations, and the clever adaptations of plants to their new homes and ecosystems.

Prickly neighbour
Adapted to arid climates where resources are scarce, Opuntia protect themselves from being eaten or trampled by herbivores with vicious spines and small prickles called glochids that drive into hair or skin.

Living together
Many plants have coevolved with other organisms and work with them in partnership. Others, such as the bee orchid (Ophrys apifera, right) resort to undercover methods to fulfil their needs, imitating female bees to attract pollinating males to their flowers.

INTRODUCTION

Section 1

ANCIENT BIOMES

THE FIRST LAND PLANTS

Roughly 475 million years ago, in one of the most significant events in the history of Earth, plants moved out of the water and evolved new traits that allowed them to survive and spread on land.

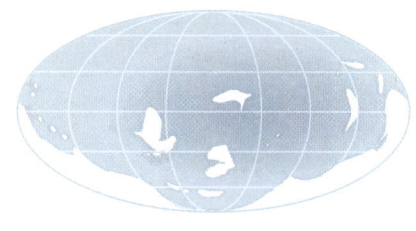

The Ordovician world
During the Ordovician Period, the majority of Earth's continental plates were located in the southern hemisphere.

ORIGIN OF LAND PLANTS

It is not known exactly when plants began to colonize land, but it is thought to have occurred sometime during the Cambrian or early Ordovician periods. The fossil record for this vitally important step in plant evolution is very limited. However, by comparing the genetic data of living species today it is possible to build a family tree and identify the closest living relatives of the first plants to make the leap to land – a group of green algae, called the Zygnematophyceae, that live in fresh water.

Green chloroplasts that facilitate photosynthesis are arranged in spirals

Living relative
The genetic and cellular structures of alga such as Spirogyra are thought to be similar to those found in the aquatic plants that gave rise to land plants.

SPORE STRUCTURE

The first direct evidence for plants on land comes from spores preserved as fossils for hundreds of millions of years. Many of these fossils have similarities to the spores of modern liverworts, including a tough outer layer composed of sporopollenin that provides a protective barrier against the harsh conditions on the terrestrial surface.

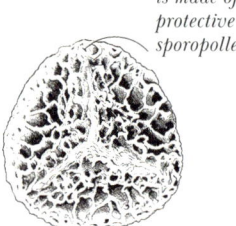

Outer layer is made of protective sporopollenin

Land-plant spore

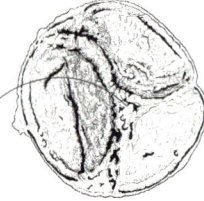

Land-plant spores often have a Y-shaped mark

Fossil spore

ADAPTING TO LIFE ON LAND

The fossil record provides limited evidence for the physical appearance and attributes of the first land plants, but comparisons with living species reveal that a key innovation was the evolution of a special mode of reproduction. In all land plants, the plant embryo is protected and nourished in some shape or form by the parent plant after fertilization. This defining feature provided the first land plants with the ability to produce and nurture reproductive spores that could then be spread and successfully establish themselves over the land surface. This trait is still present in plants today – from mosses and liverworts to conifers and flowering plants.

Living comparisons
Modern liverworts are a diverse group of plants that develop flat ribbonlike or leafy bodies from which complex reproductive structures form on stalks. They also produce spores that are transported by wind and water.

VASCULAR PLANTS FORM

Fossils from the Silurian Period provide the first glimpse of early plant form and function, charting the rise of plants with specialized tissues for transporting water internally.

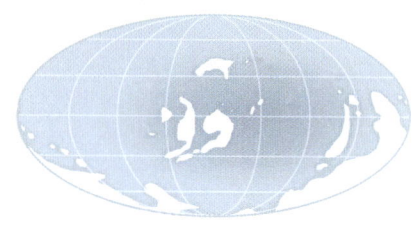

Silurian world map
By 430 million years ago the climate had become warmer and the landmasses that were to become Euramerica had started to collide.

SURVIVING ON LAND

Many Silurian plant fossils are assigned to the genus *Cooksonia*, which were only a few millimetres or centimetres high, and lacked leaves and roots. These species had evolved many adaptations for life on land. One of these was a waterproof outer layer (cuticle) to prevent drying out and to provide protection against pathogens and UV damage. However, the cuticle limited the plants' ability to absorb carbon dioxide (CO_2) from the atmosphere, an essential gas for photosynthesis. To absorb CO_2, plants evolved pores, called stomata.

Stomata
The outer, epidermal, layer of plants is covered in protective cuticle. To allow the plant to control water and gas exchange with the environment, special guard cells can open and close pores.

Central pore is open to take in carbon dioxide

Closed stoma prevents loss of moisture

Epidermal cell covered in cuticle

Guard cell opens and closes stoma

Stoma open **Stoma closed**

Reproductive structure (sporangium) at the end of each branch

The sporangia contain thousands of spores

Branching increases the number of sporangia

Stems have vascular tissue (see right)

Cooksonia
In living plants, stomata are mainly found on the leaves. Being leafless, it was the stems of Cooksonia that did the work of photosynthesising and had the stomata.

THE SILURIAN PERIOD

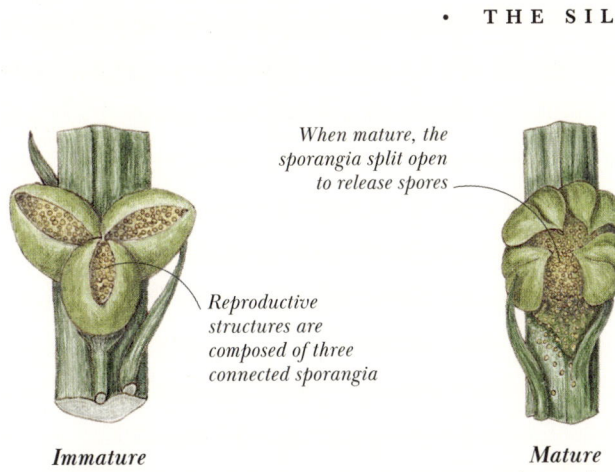

When mature, the sporangia split open to release spores

Reproductive structures are composed of three connected sporangia

Immature sporangium

Mature sporangium

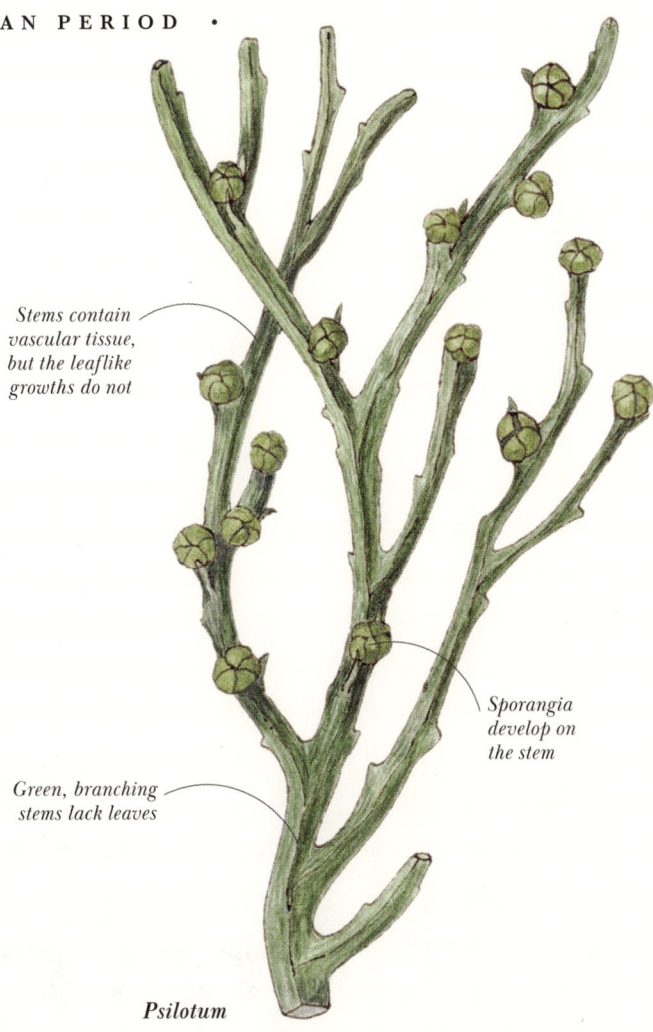

Stems contain vascular tissue, but the leaflike growths do not

Sporangia develop on the stem

Green, branching stems lack leaves

Psilotum

A GLIMPSE OF THE PAST

The vast majority of living plants have leaves for capturing sunlight and roots to anchor them and take up water and nutrients. The first land plants lacked both. An example of a living species that gives us a good idea of what these early land plants might have been like is *Psilotum*. Like *Cooksonia*, it has no leaves or true roots and, in the same way as Silurian plants, it relies primarily on spores to reproduce. This form of reproduction tied early plants to wet conditions as the fertilization process required water to bring male and female gametes together.

Whisk fern
Psilotum nudum, commonly called the whisk fern, provides a modern example of what the Silurian flora might have looked like, with its leafless, photosynthetic stems.

VASCULAR TISSUES

As plants became larger and more complex, they needed to evolve a transport system for food and water. This resulted in two types of vascular tissue: woody xylem, is made of dead cells and transports water; phloem is made of living cells, and transports food. Xylem also provides structural support.

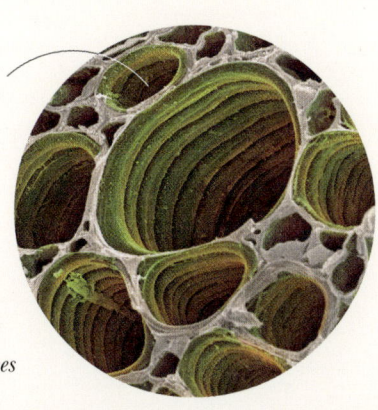

Xylem cells are dead and hollow

Xylem tissue
Vascular cell walls are reinforced with a substance called lignin, which prevents them from collapsing and gives the plant structural support.

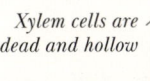

Ladderlike thickenings provide good support

Hooplike thickenings are found in the first-formed xylem (protoxylem)

Structural flexibility
Xylem cells are reinforced in different ways in different plants. These differences are used to help assign species to different groups of vascular plants.

Reproductive sporangia are grouped together to form a fertile zone

FIRST COMPLEX ECOSYSTEMS

During the Early Devonian Period, roughly 419–393 million years ago, plants continued to diversify. The fossils of these ancient ecosystems are found around the world and provide the first evidence of primitive leaves and roots.

GREENING THE CONTINENTS

The first complex ecosystems evolved on land and were now underpinned by vascular plants. As in the Silurian, most of the plants in these ecosystems would have looked quite alien today with their naked green stems for capturing light and only small hairlike cells, called rhizoids, to anchor them. However, among these species we can also find evidence that plants were innovating. As ecosystems became more crowded, plants evolved two game-changing innovations to survive: they grew leaves to increase their capacity to capture sunlight, and developed roots to search for water and nutrients in thin soils.

A particularly rich fossil deposit in Scotland, called the Rhynie chert, captures in its sedimentary rock not only plants but also arthropods and beneficial fungi – a reminder that plants were not evolving in isolation.

The fossil record
The plants at this time were all relatively low growing. Baragwanathia stood only 28cm (1ft) high. Pertica quadrifaria, found in what is now North America, was up to 1.8m (6ft) tall.

Early Devonian ecosystems
Early lycophytes were small herbaceous plants and had early microphyll leaves. They grew in stands of intertwined stems alongside the leafless Trimerophytes.

Baragwanathia
Lycophyte

Sawdonia
Lycophyte

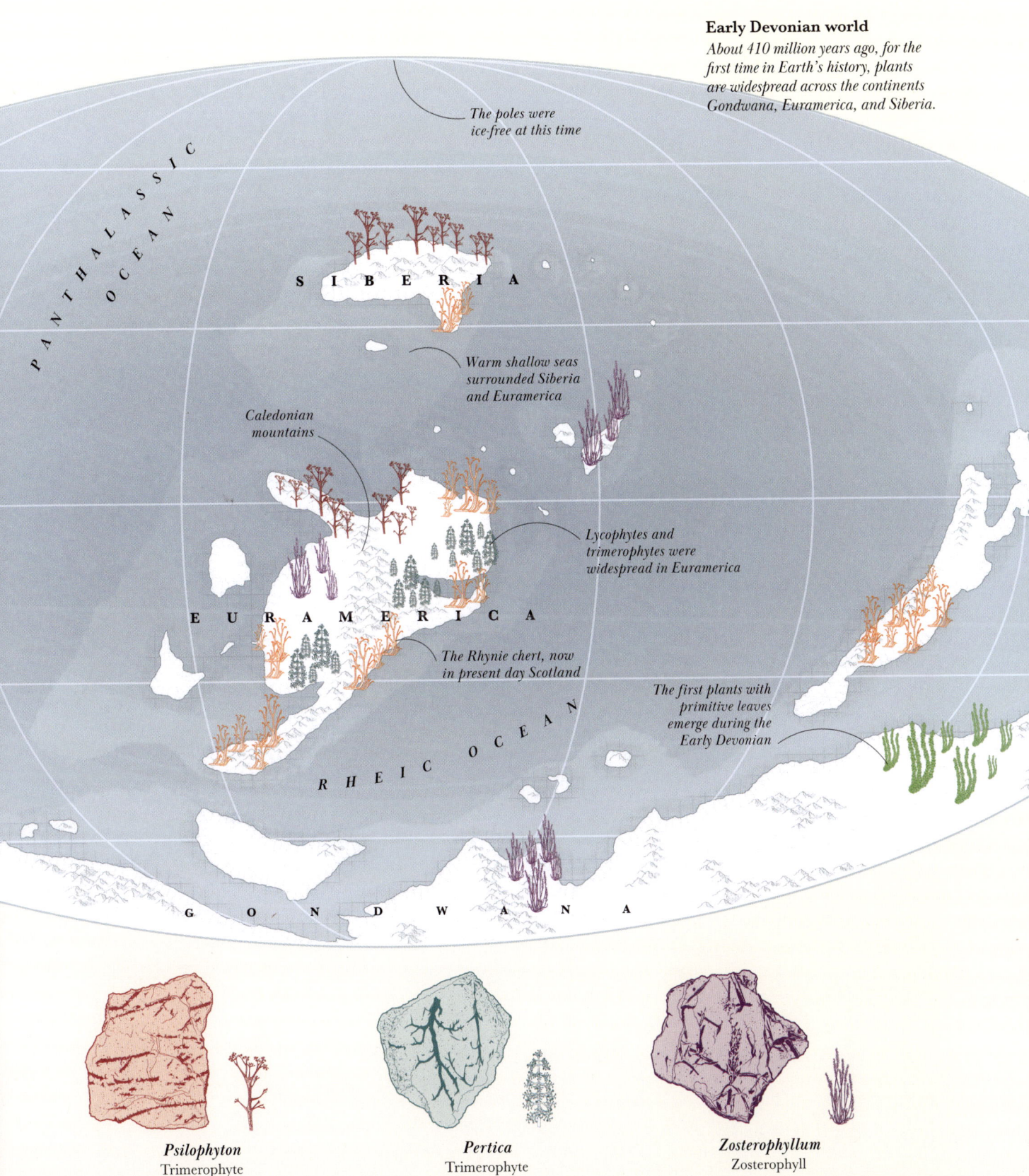

Lycopodium clavatum

Some living lycophytes, such as this clubmoss (Lycopodium clavatum), look very similar to the first leafy species from the Early Devonian. Leafy shoots spread across the ground, producing underground roots and aerial fertile strobili.

Reproductive structures called strobili release spores, which are scattered by the wind

Scaly leaves cover the stem

The plant spreads from horizontal leafy shoots called rhizomes

Roots develop from the rhizome in search of water and nutrients

THE ORIGIN OF LEAVES

With their broad, flat surface, leaves are highly effective at capturing the sunlight needed for photosynthesis. They are undeniably one of the most iconic and diverse features of plants today and first became widespread in the Early Devonian. Before they had leaves, plants relied on simple green stems to carry out photosynthesis. By developing leaves, plants dramatically increased their surface area and hence their capacity to capture light and fuel the necessary processes for growth and reproduction. Leaves evolved multiple times in land plants, but the first group of plants with leaves were the lycophytes, specifically those from the group called the Drepanophycales. They developed leaves called microphylls, with a characteristic single, unbranched mid-vein. Their fossils are known from the very end of the Silurian and become widespread in the Early Devonian. Leafy shoots and underground roots were ideal for Drepanophycales to thrive in new environments and their fossils are found across the world.

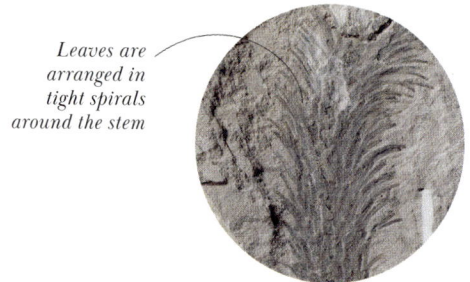

Leaves are arranged in tight spirals around the stem

Baragwanathia

The first fossils of one of the earliest leafy plants, Baragwanathia, were found in Australia. Its shoots were covered in hundreds of small needlelike leaves. It is a lycophyte and member of the Drepanophycales.

THE EARLY DEVONIAN PERIOD

THE ORIGIN OF ROOTS

Before roots evolved, plants relied on a smaller anchoring system of rhizoids, which non-vascular plants, such as liverworts, still use. Roots were one of the key innovations of vascular plants, and like leaves they first evolved in lycophytes related to *Baragwanathia*. The evolution of roots allowed plants to enormously increase their anchorage and their ability to access water and nutrients. They were therefore vital for allowing plants to populate new, more arid areas.

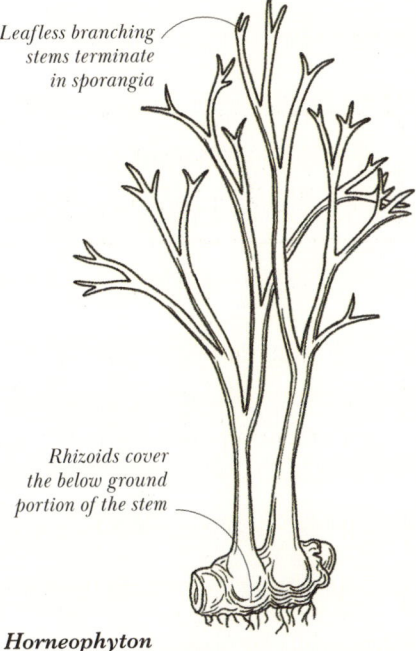

Leafless branching stems terminate in sporangia

Rhizoids cover the below ground portion of the stem

Horneophyton
This extinct Early Devonian species lacked roots and produced huge numbers of rhizoids for anchorage and for water and nutrient uptake.

Umbrella liverwort
Liverworts produce a flat green body called a thallus for capturing light and are anchored to the substrate by their rhizoids.

Plant not differentiated into true leaves and stems

Rhizoids form on the underside of the plant

SYMBIOSIS

The first direct evidence of complex early ecosystems on land is preserved from the Early Devonian Period. An example of one of these ecosystems can be found in the Scottish fossil site called the Rhynie chert. Fossils in the Rhynie chert show such exceptional levels of preservation that even minute organisms can be identified. The Rhynie chert has therefore been key for documenting early evidence of symbiosis. Symbiosis is an interaction between two different organisms living in close physical association. Today over 90 per cent of plants form beneficial symbioses with fungi, and fossils in the Rhynie chert demonstrate that such partnerships date back over 407 million years. The Rhynie chert also preserves the earliest certain evidence of fossil lichens, another example of symbiosis.

Lichen
Widespread in the Early Devonian, lichens consist of a symbiotic relationship between a fungus and a photosynthetic organism.

Lichens grow on rocks, tree trunks, and plant stems

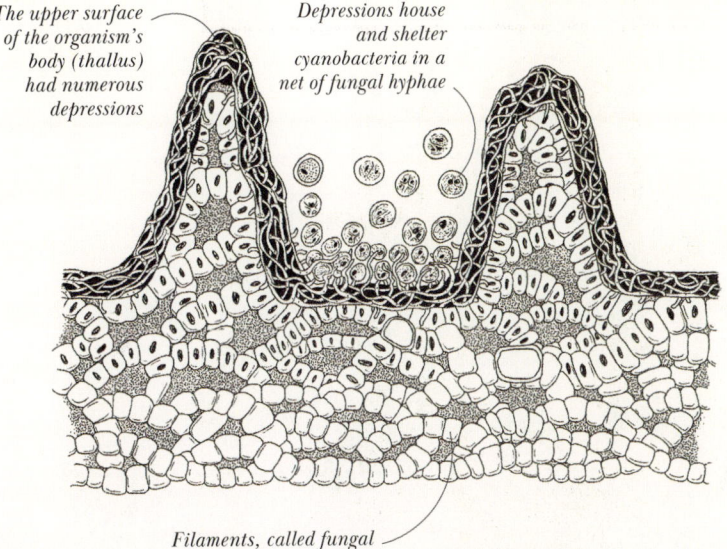

The upper surface of the organism's body (thallus) had numerous depressions

Depressions house and shelter cyanobacteria in a net of fungal hyphae

Filaments, called fungal hyphae, absorb nutrients

Winfrenatia reticulata
This extinct lichen was a symbiosis between cyanobacteria and a fungus. The cyanobacteria captured sunlight for photosynthesis and created sugars. The fungus received sugars in return for providing a sheltered environment.

Early forest inhabitants
Calamophyton was an early fernlike plant in the Cladoxylopsids. These trees formed some of the earliest forests in what is now Europe and North America.

Crown composed of many fine branches that captured light

FIRST TRUE FORESTS

In the mid-Late Devonian, roughly 390–360 million years ago, multiple separate groups of plants evolved into trees. These trees formed the first forests, a biome that has persisted from the Devonian until the present.

TREES EVOLVE

As competition for water, nutrients, and sunlight intensified, plants greatly increased in size and evolved into trees. It was the advent of wood that facilitated this. Its water-conducting xylem tissue (see p.19) provided a rigid structure and reinforced the trunks, allowing plants to grow taller. Leaves could now grow far above the forest floor and large roots extended multiple metres into the ground to find greater reserves of water and nutrients.

Trees did not just have one evolutionary origin but evolved at least three separate times: in the lycophytes, in the ferns, and in ancestors of the seed plants, the progymnosperms.

Each group was adapted to a different environment. For example, lycophytes that grew up to 3–8 m (10–26 ft) in height and early fernlike plants, called the Cladoxylopsids, thrived in wetter, swampy environments. Giant *Archaeopteris* trees, which could reach 20–30 m (65–98 ft), were adapted to seasonally drier areas.

THE LATE DEVONIAN PERIOD

The Late Devonian world
About 360 million years ago, forests covered vast swaths of the continental surface for the first time. These forests had a long-term effect on the climate.

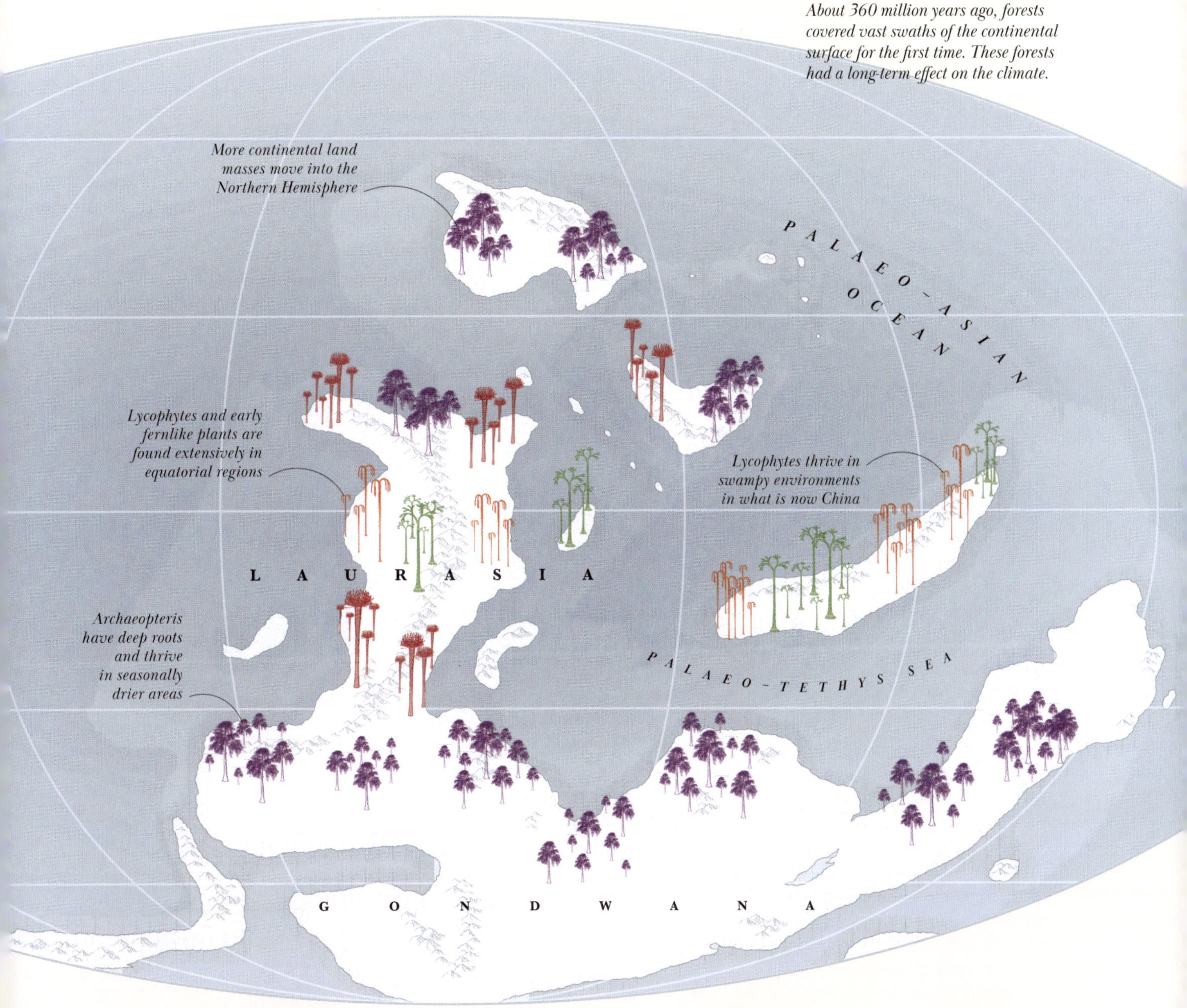

More continental land masses move into the Northern Hemisphere

Lycophytes and early fernlike plants are found extensively in equatorial regions

Lycophytes thrive in swampy environments in what is now China

Archaeopteris have deep roots and thrive in seasonally drier areas

PALAEO-ASIAN OCEAN

LAURASIA

PALAEO-TETHYS SEA

GONDWANA

The fossil record
Land plants increased in size and Late Devonian fossils show the increase in water conducting cells (xylem), which provided the internal support to keep taller plants upright.

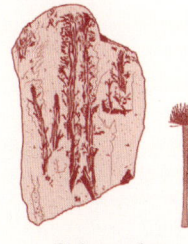

Calamophyton
Fernlike plant

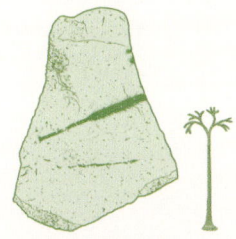

Chamaedendron
Cladoxylopsid

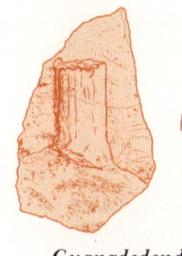

Guangdedendron
Lycophyte

Archaeopteris
Progymnosperm

THE INFLUENCE OF TREES

Trees fundamentally transformed Earth. As extensive forests spread, trees locked up carbon in their woody trunks and gradually drew down CO_2 from the atmosphere, leading to global cooling. Their deep roots helped create and stabilize soils, which in turn impacted river systems. Earlier rivers were composed of many ribbonlike streams spreading out across flood plains. However, the stabilizing influence of trees reinforced river banks giving rise to meandering river channels.

Gnarled and twisted trunk allowed mosses, lichens, and epiphytes to grow directly on the bark

Highly branching photosynthetic branches formed the crown of the fernlike plant Calamophyton

Fernlike plant

Chamaedendron was a small lycopsid tree with reproductive cones, called strobili

Clubmoss tree

Aneurophyton would have looked like woody shrubs or large vines

Progymnosperm

Aged tree trunk
Trees can have remarkable longevity, with the oldest being thousands of years old. This makes them longer-term hubs of biodiversity, supporting complex ecosystems that grow up around them.

Extinct woody plants
The lycophytes, progymnosperms, and fernlike plants all contributed to the diversity of the new forests, finding their own niches. Tree species in these groups gradually went extinct and were replaced by seed plants, such as conifers and flowering plants.

• THE LATE DEVONIAN PERIOD •

THE FIRST SEED PLANTS

The Late Devonian provides the first fossil evidence for seeds. This represents the starting point for the remarkable diversity of living seed plants that include conifers, cycads, ginkgo, and flowering plants. Seeds and pollen were crucial innovations. Seeds provide protection and nourishment for the developing plant embryo, and both seeds and pollen moved plants away from their reliance on water for reproduction (see p.19) and enabled them to diversify and spread into drier environments.

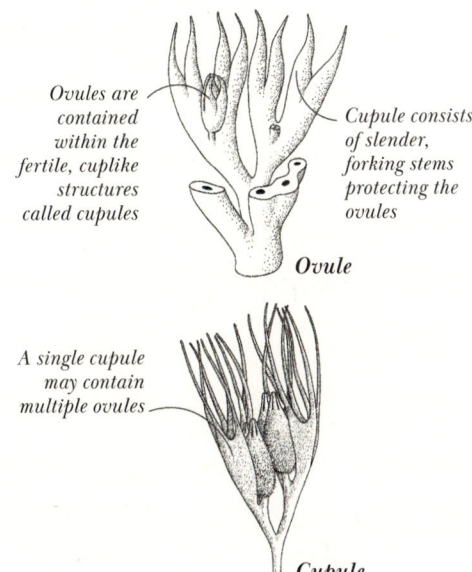

Ovules are contained within the fertile, cuplike structures called cupules

Cupule consists of slender, forking stems protecting the ovules

Ovule

A single cupule may contain multiple ovules

Cupule

Pollen grains contain the male sex cells of the plant

The pollen tube grows out of apertures to deliver male gametes to the ovule

Wind-blown pollen
These microscopic pollen grains are the male gametophytes. The tiny size of pollen grains makes them ideal to be transported by the wind.

Reproductive parts
One of the earliest-known seed plants is Elkinsia. Fossils of this have been found in Late Devonian deposits in West Virginia, US. Its seed-bearing ovules were completely enclosed in structures called cupules.

Root types
Deep woody roots are ideal for pushing deep into the ground in search of water, whereas highly branched roots close to the surface are better suited for rapidly absorbing nutrients in waterlogged conditions.

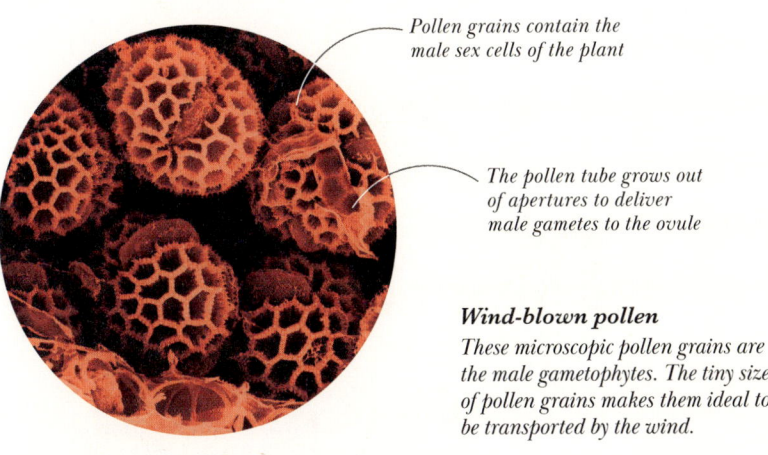

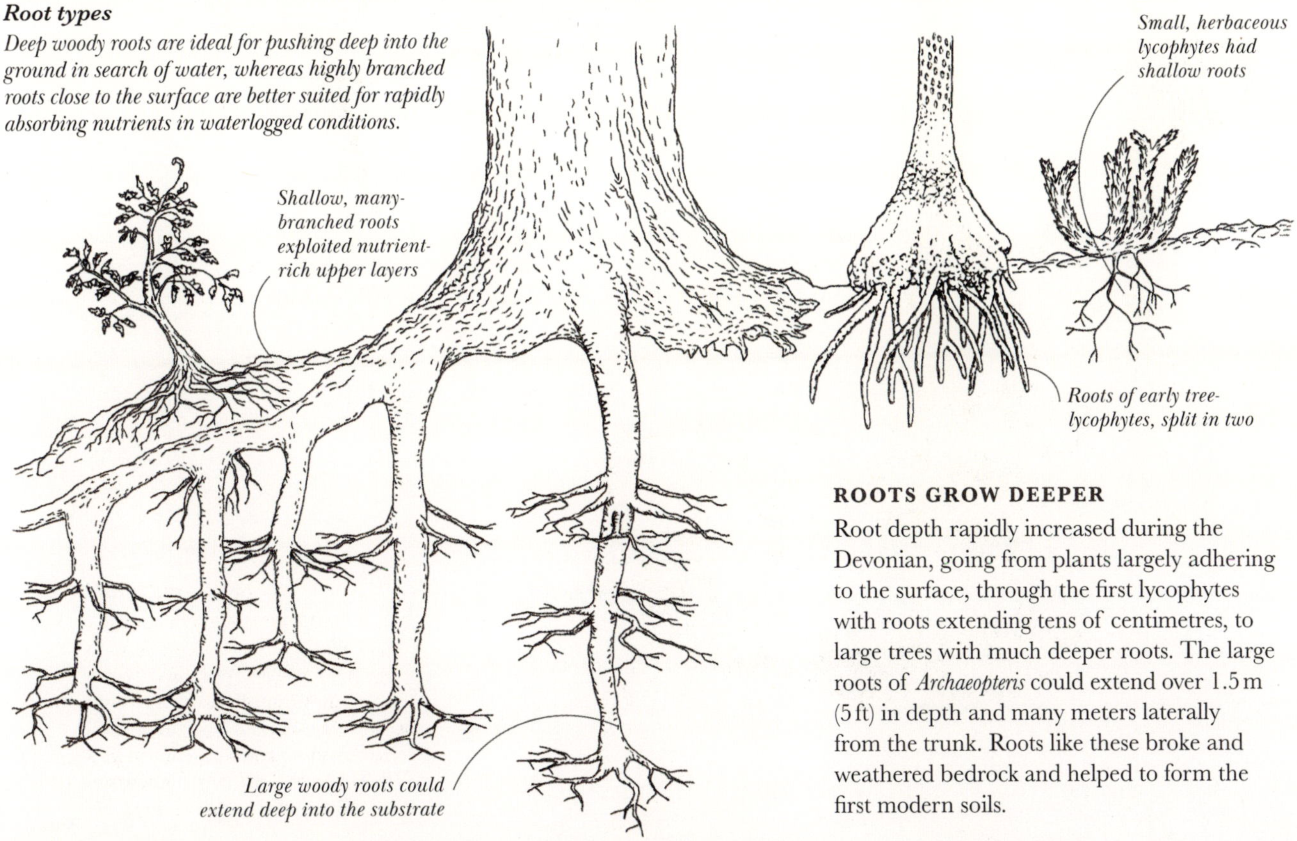

Shallow, many-branched roots exploited nutrient-rich upper layers

Small, herbaceous lycophytes had shallow roots

Roots of early tree-lycophytes, split in two

Large woody roots could extend deep into the substrate

ROOTS GROW DEEPER

Root depth rapidly increased during the Devonian, going from plants largely adhering to the surface, through the first lycophytes with roots extending tens of centimetres, to large trees with much deeper roots. The large roots of *Archaeopteris* could extend over 1.5 m (5 ft) in depth and many meters laterally from the trunk. Roots like these broke and weathered bedrock and helped to form the first modern soils.

Swamp forests
Coal swamp forests spanned vast areas of the tropics, forming a major band from the west across Euramerica to the islands in the east.

Scale trees are named for their scaly bark

COAL FORESTS EVOLVE

During the Carboniferous Period, 360–300 million years ago, lush swamps of large trees covered the equatorial and tropical regions. Over time their remains built up to form coal deposits that are found across the world today.

FOREST GIANTS

Spore plants dominated the Carboniferous coal swamp forests. Whereas most ferns and clubmosses today are small herbs, the ancient forests included towering clubmosses that reached heights of about 50 m (165 ft), and colossal horsetails that grew up to 20 m (65 ft) tall. Below the canopy, coal forests were also home to seed plants with fernlike foliage – known as seed ferns – which took up residence on the forest floor.

Life in the coal swamp forests was hot and humid, and plants developed adaptations – such as broad, supportive root systems – that allowed them to thrive in the waterlogged conditions. However, outside these equatorial swamps, Earth was gradually cooling to temperatures similar to those of today. This cooling process resulted in the formation of a major ice cap in the Southern Hemisphere, and lead to the expansion of arid biomes, in which plant life became more sparse.

The fossil record
Fossil finds reveal that adaptations such as scaly bark, spikey leaves, and ferny fronds had become typical features of plants in the Carboniferous.

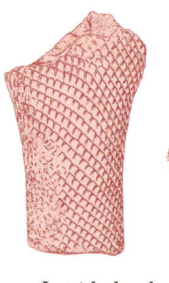

Lepidodendron
Tree clubmoss

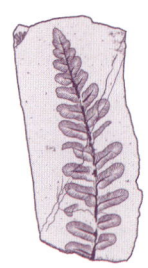

Neuropteris
Seed fern

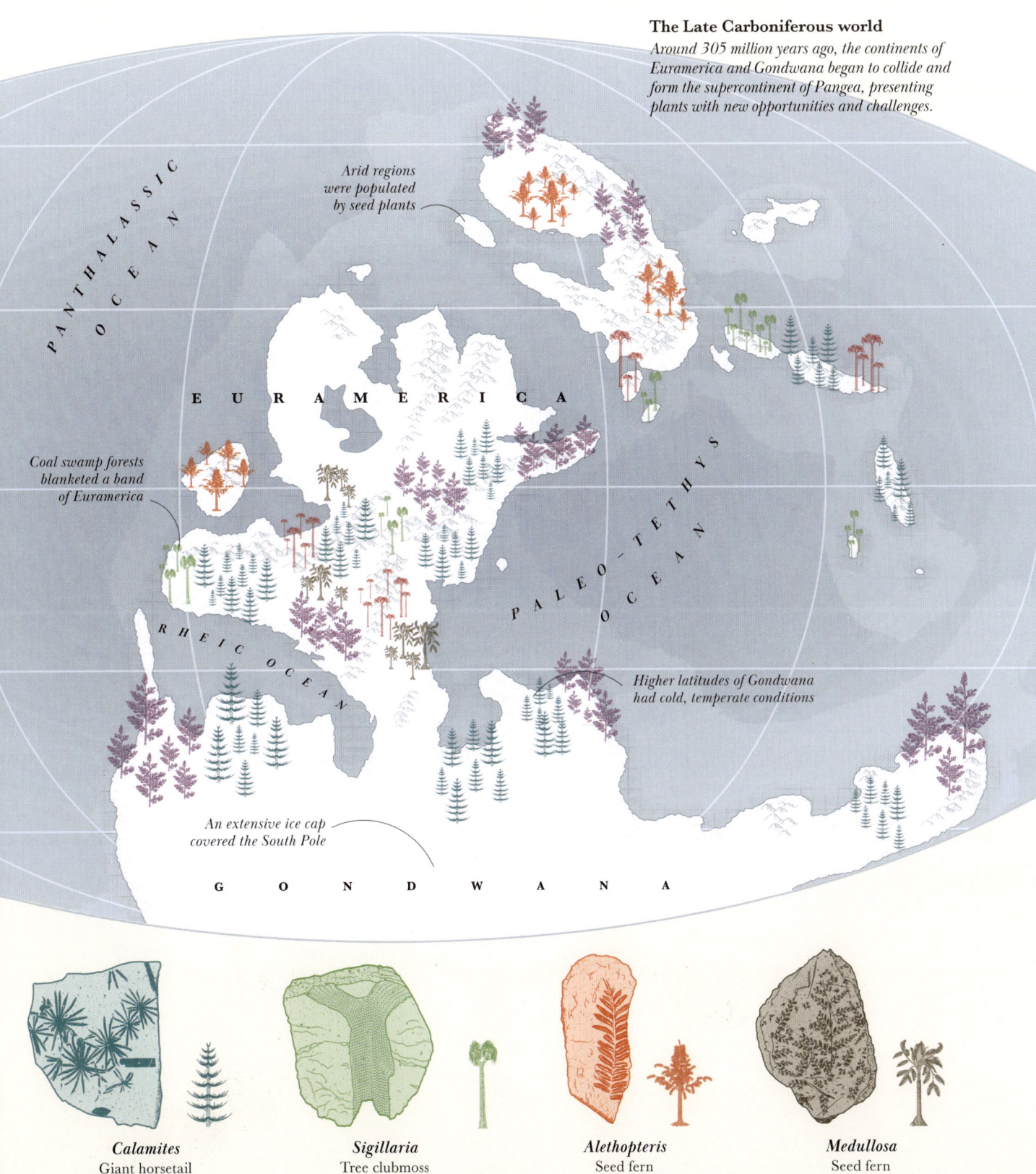

ADAPTATIONS TO SWAMP LIFE

Growth in waterlogged ground posed two challenges for plants: gaining secure anchorage and supplying their roots with sufficient oxygen. The tree clubmosses developed extensive woody roots, called *Stigmaria*, which were densely covered in smaller rootlets – sometimes more than 25,000 per metre (over 7,600 per foot). This extensive rooting system helped stabilize these giant trees. To supply oxygen to underground roots, tree-sized horsetails had a large cavity in their rhizomes, the underground portion of the stem, allowing them to move oxygen around the plant. Modern horsetails that live in water use a similar adaptation.

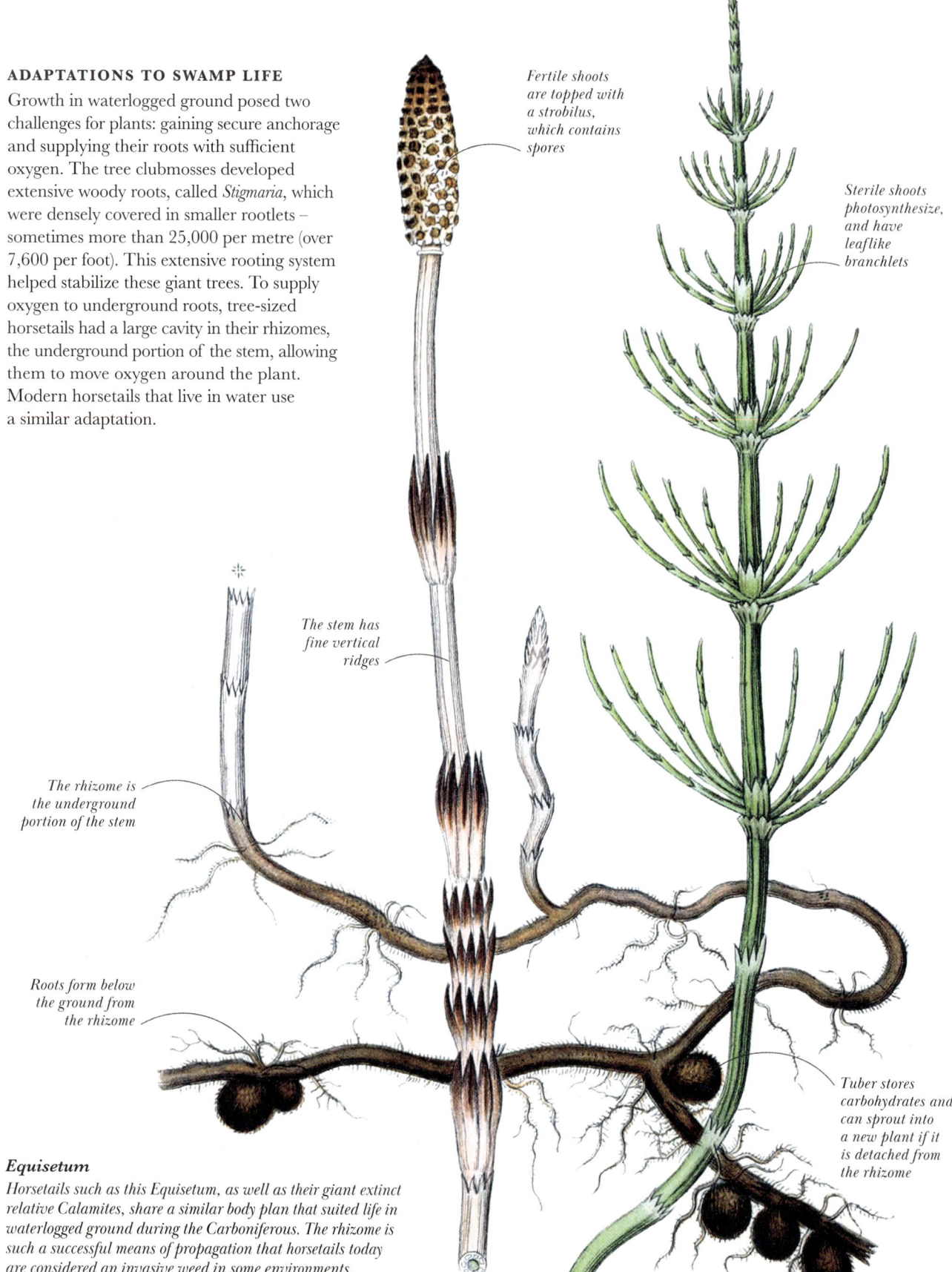

Fertile shoots are topped with a strobilus, which contains spores

Sterile shoots photosynthesize, and have leaflike branchlets

The stem has fine vertical ridges

The rhizome is the underground portion of the stem

Roots form below the ground from the rhizome

Tuber stores carbohydrates and can sprout into a new plant if it is detached from the rhizome

Equisetum
Horsetails such as this Equisetum, as well as their giant extinct relative Calamites, share a similar body plan that suited life in waterlogged ground during the Carboniferous. The rhizome is such a successful means of propagation that horsetails today are considered an invasive weed in some environments.

THE CARBONIFEROUS PERIOD

Extinct horsetails
Horsetails were highly diverse in the Carboniferous. Some were small and herbaceous while others were very tall plants. Calamites could reach heights of 20m (66ft). Many of these plants were preserved within coal deposits. Annularia is the name given to the fossil leaves of Calamites.

Leaves grew in rings round the stem
Annularia

Fossil stems had characteristic ridges
Equisetites

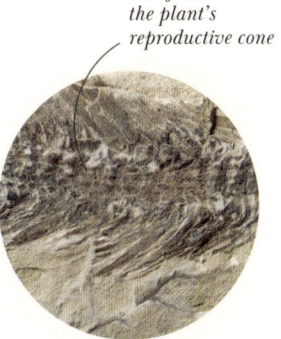
This fossil shows the plant's reproductive cone
Calamites

PROTECTING NEW FRONDS

Fiddleheads, the tightly coiled young fronds of ferns before they unfurl, first appeared in the Carboniferous. These tight coils protect the fragile growing tips from extreme temperatures, desiccation, UV radiation, and pathogens. As ferns diversified, growing to tree-sized plants with longer fronds that took longer to develop, this protection of the vulnerable leaves became essential.

Fiddleheads
In temperate biomes, fiddleheads unfurl during spring as the temperature warms and light becomes more available.

Fiddlehead protects immature leaves

From bud to frond
Tiny buds grow from the rhizome and form coiled fiddleheads. Young leaves develop in the fiddlehead before expanding into a full frond.

The rhizomes may persist below ground for many years

CLIMBING PLANTS EVOLVE

As competition for light and space on the forest floor increased, plants evolved new strategies. Coal swamp forests provide the first evidence for climbing plants, for example; seed ferns evolved hooks, tendrils, and twining habits to climb from the ground. Some plants evolved to live on larger tree ferns – as epiphytes. Life above the ground opened up new possibilities, but without deep roots epiphytes needed to rely on rain for water.

Botryopteris forensis
This fern is considered to be one of the earliest examples of an epiphyte. It grew on the trunk of a larger fern called Psaronius.

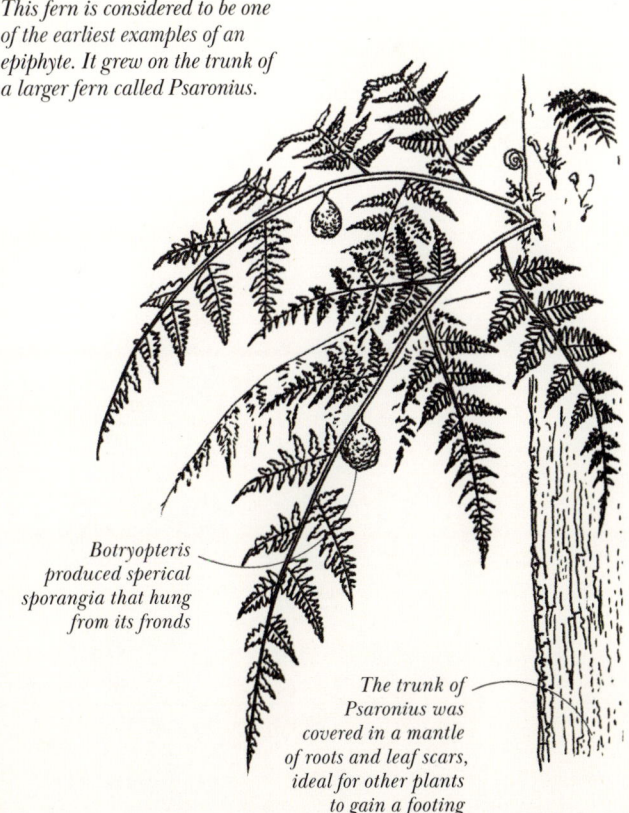

Botryopteris produced sperical sporangia that hung from its fronds

The trunk of Psaronius was covered in a mantle of roots and leaf scars, ideal for other plants to gain a footing

Voltzialean conifers

Found in northern and southern regions of Pangea, Voltzialean conifers are an extinct group that share specific features – such as their habit, scaly leaves, and cone structure – with a range of modern conifer species.

Regularly spaced branches were covered in scalelike leaves

THE RISE OF CONIFERS

During the Permian Period about 300–250 million years ago, warm, increasingly arid conditions on the supercontinent Pangea supported the expansion of seed plants and the rise of drought-tolerant conifers.

DRYING WORLD

The coming together of the continental plates that formed the supercontinent Pangea had a significant impact on Earth's climate. The wet, humid coal swamp forests of the Carboniferous collapsed and gave way to more arid conditions. Further south, plants began to colonize new tracts of land that became available as the polar ice cap melted.

Seed plants diversified in this warmer ice-free world, and evolved a suite of adaptations that helped them to thrive in different and more varied environments. Crucially, their seeds protected the embryo and provided nourishment, which allowed plants to tolerate harsh conditions and germinate only when circumstances became favourable. They also changed their foliage: at high altitudes large-leaved species captured precious light during long polar summers. In drier regions, plants that produced needle- or scalelike leaves to reduce water loss were able to succeed in harsh, arid environments.

One group of seed plants in particular – conifers – diversified extensively during this period. The conifers established themselves as a major component of Earth's flora, and remain a key element of modern ecosystems today.

The fossil record

Fossils from the Permian Period show how plants adapted their foliage to cope with different conditions.

THE PERMIAN PERIOD

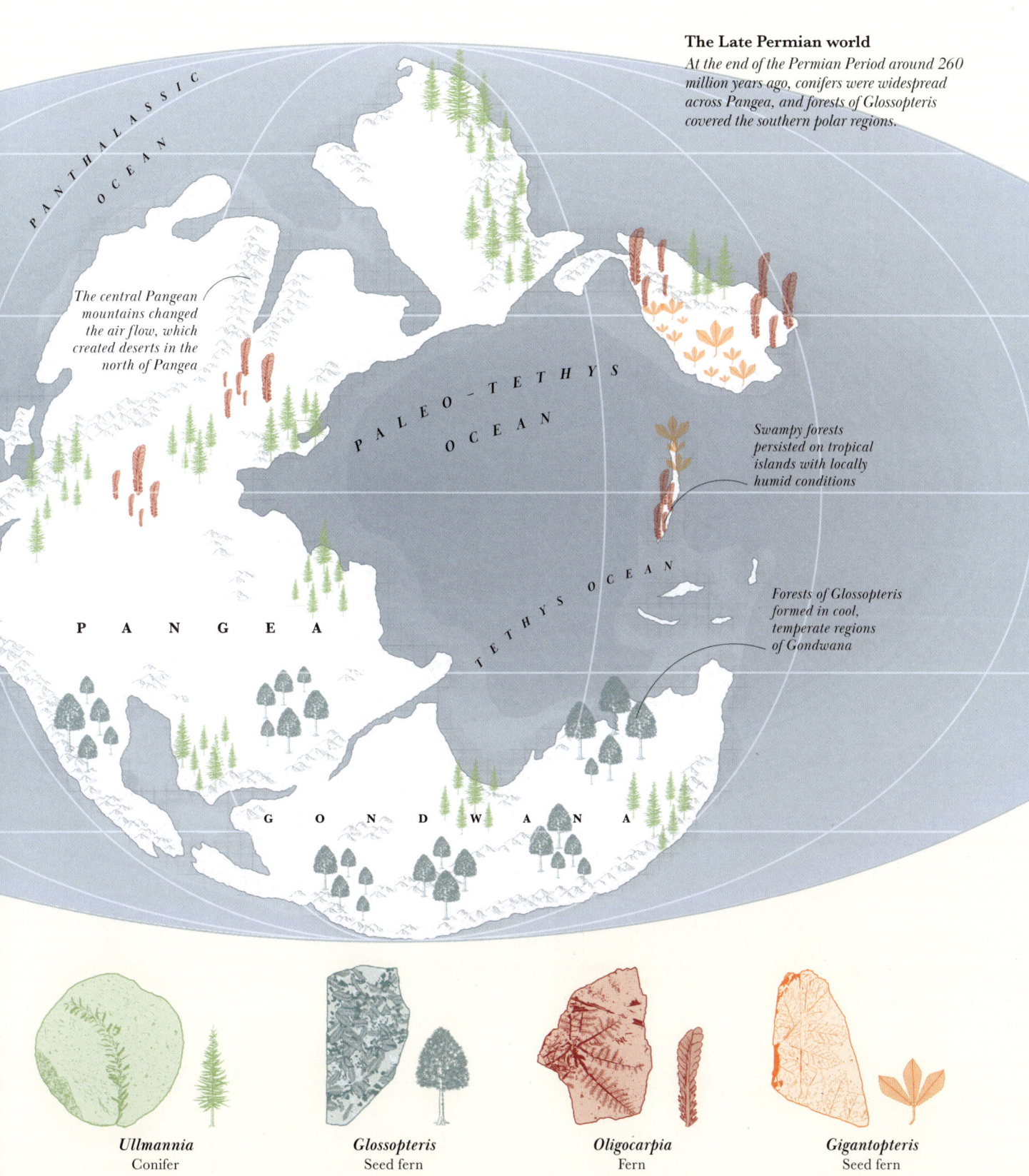

The Late Permian world
At the end of the Permian Period around 260 million years ago, conifers were widespread across Pangea, and forests of Glossopteris covered the southern polar regions.

The central Pangean mountains changed the air flow, which created deserts in the north of Pangea

Swampy forests persisted on tropical islands with locally humid conditions

Forests of Glossopteris formed in cool, temperate regions of Gondwana

Ullmannia
Conifer

Glossopteris
Seed fern

Oligocarpia
Fern

Gigantopteris
Seed fern

FIRST MODERN CONIFERS

Conifers flourished during the Permian and diversified into the major groups that are still alive today. The warmer and more arid biomes favoured these plants with their scalelike or needlelike leaves, thick waxy cuticles, and sunken stomata, all of which helped to limit water loss. The plants were also well adapted to the risk of wildfires or drought and some plants had the ability to drop branchlets or even whole branch systems. A new diversity of cones is also recorded in Permian fossils.

Ernestiodendron filiciforme
The conifer Ernestiodendron was a widespread, tall forest tree. Its habit was similar to the living Norfolk Island pine, but with longer, curved leaves.

Dense foliage with curved leaves

Permian conifer

Norfolk Island pine
Although only distantly related to its Permian ancestors, Araucaria heterophylla still preserves many of the characteristics present in the Permian voltzialean conifers, such as the dense covering of scaly leaves, regular arrangement of branchlets, and the aggregation of fertile parts into cones.

Leaves arranged in tight spirals around branchlet

Small leaves with a thick, waxy covering

Large mature female cone is composed of spirally arranged scales

Immature cones are green and can photosynthesize

Immature cone   **Mature cone**

• THE PERMIAN PERIOD •

DECIDUOUS LEAVES

A warmer world with no ice at the poles meant plants thrived at high latitudes. The southern pole was cooler and damper than continental interiors but would have experienced seasonal extremes. One way that plants, such as *Glossopteris*, adapted to these conditions was the evolution of deciduousness. Large leaves were ideal for capturing light in the long summer but were shed in the autumn to avoid water loss and damage over the darker polar winter.

Glossopteris
The leaves of this seed fern grew on woody branches and investigation of its fossilized tree rings in the wood indicates that plants experienced rapid growth spurts during the spring and summer months.

Leaves had a characteristic complex vein network

Leaves attached to woody stems in a spiral pattern, to maximise light capture

Large tongue-shaped leaves

Leaf litter
Glossopteris fossils are found extensively in India, Australia, Antarctica, Africa, and South America, once part of the ancient continent of Gondwana. Leaf fossils occur in carpets, reflecting the deciduous strategy of the plant.

Deciduous leaves
Many plants shed their leaves annually during the autumn to avoid damage and water loss in harsh winter conditions. Before being shed, leaves change colour as the green chlorophyll is broken down and other pigments become visible, giving leaves yellow, orange, and red colours.

Only some conifers shed their leaves

Larch

Many broadleaf trees lose their leaves

Oak

Bright golden yellow autumn colour

Ginkgo

SEEDS

A key adaptation of conifers was the evolution of seed dormancy. This is when after fertilization the plant embryo remains in a state of suspended development encased within the protected seed. When conditions become favourable germination occurs, kick-started by nutritious tissues in the seed. Seed dormancy was an ideal innovation in response to the harsh environmental conditions found in the Permian.

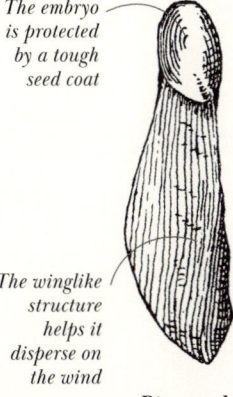

The embryo is protected by a tough seed coat

The winglike structure helps it disperse on the wind

Pine seed

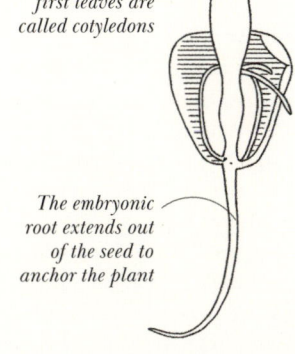

The embryonic first leaves are called cotyledons

The embryonic root extends out of the seed to anchor the plant

Inside a seed

The right time to germinate
Seeds are dispersed from their parent plants, for example by wind or by animals, and only when the conditions are right do the seeds germinate. The seeds first produce leaves and an embryonic root to help the growing plant become established.

Pioneer plants

Pleuromeia, a treelike spore plant that could grow to 2 m (6.5 ft) tall, greatly expanded its range in the early Triassic, while Dicroidium, with its characteristic forked leaves, became a stalwart of the Southern Hemisphere.

Pleuromeia was topped with a reproductive spore cone

FERNS AND SEED PLANTS DIVERSIFY

The Triassic Period, about 250–200 million years ago, was an era of significant change. It started with a global mass extinction and arid conditions, but plants adapted and bounced back with a resurgence of both spore and seed plants.

COLLAPSE AND RENEWAL

The transition from the Permian to the Triassic saw the most severe mass extinction in Earth's history – estimates suggest that more than 57 per cent of all biological families were lost. Plants were affected by the widespread collapse of their ecosystems. However, in the hot, arid world that followed, some groups not only survived, but thrived.

Plants are often among the first organisms to rebound after major upheaval. While the world was still reeling, spore plants surged in abundance, especially *Pleuromeia*, a relative of the giant clubmosses that once dominated many landscapes.

After this initial spike, other biomes gradually recovered. Tropical regions once again began to support conifers and cycads, and temperate biomes saw mosses, ferns, and seed plants thrive.

The Triassic also saw the evolution of two distinct forms of seed plant foliage: the forked leaves of *Dicroidium* and the fan-shaped leaves of the Ginkgoales. Both forms of foliage became enduring features of prehistoric forests.

The fossil record

Triassic fossils reveal that a diverse flora evolved as ecosystems revived, with conifers, cycads, seed ferns, ginkgos, ferns, and mosses shaping the recovering forests.

Osmundales
Fern

Voltzia
Conifer

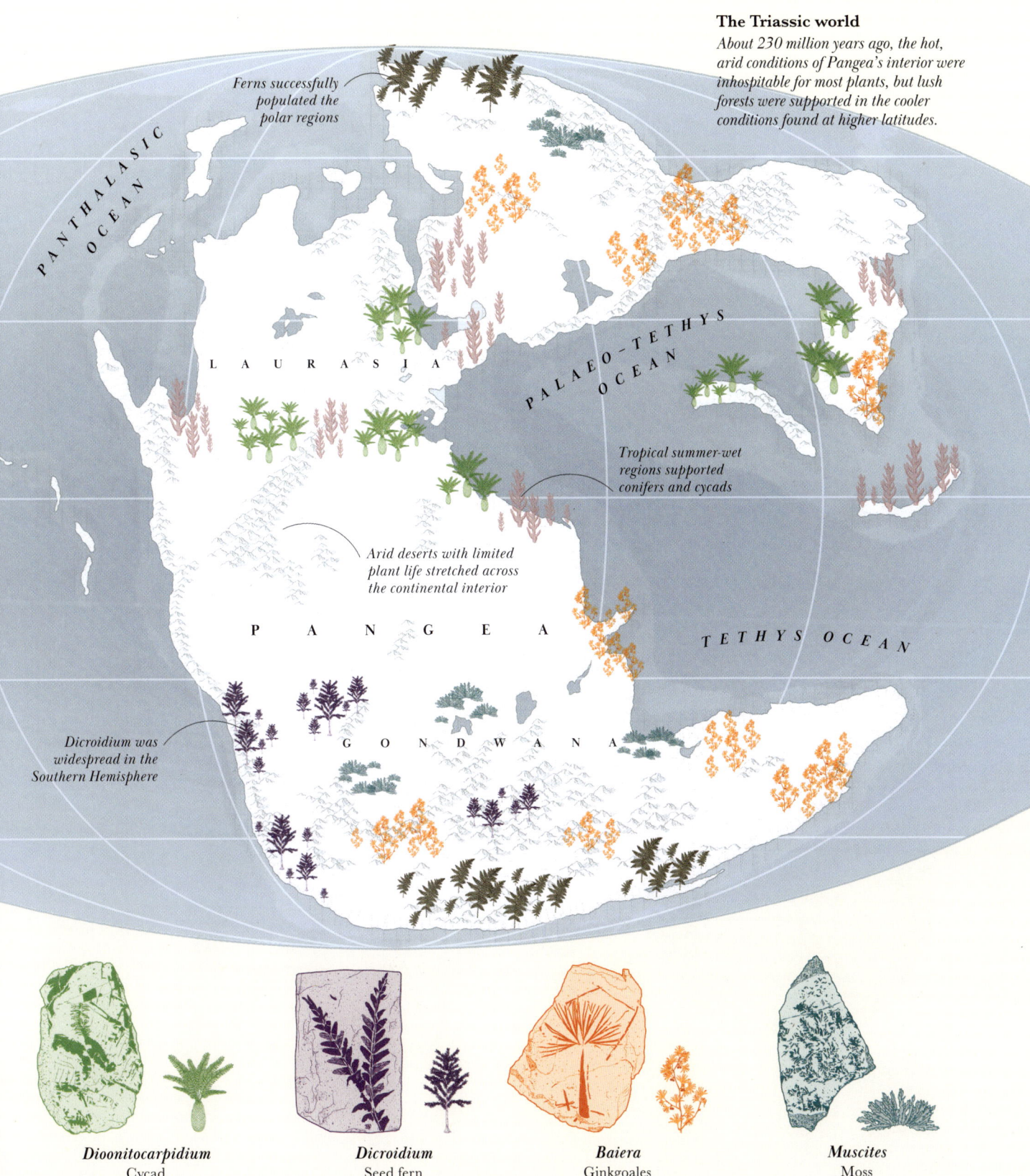

SURVIVING MASS EXTINCTION

The Triassic opened with ecosystem collapse: widespread deforestation, wildfires, flooding, spikes in global temperature, and dangerously high levels of UV radiation. Yet plants weathered this shock and eventually rebounded. Plants possess many adaptations to survive extremes, such as subterranean structures, the ability to regenerate, and, crucially, resistant spores and seeds that could germinate when conditions improved.

Robust trunklike stem with leaf scars

Osmunda fossil

Modern fern fronds closely resemble their fossil ancestors

Growing back
Ferns have a remarkable ability to bounce back after catastrophe. The Osmundales survived the end-Permian extinction event and still exists today. Spores can be widely dispersed by wind or water, finding new areas to colonize, such as this lava flow on Réunion.

Fan-shaped leaves with branching, nearly parallel veins

Fleshy outer surface of seed

Leaves are clustered on short shoots

Ginkgo biloba
One of the world's most distinctive seed-plant trees, Ginkgo biloba is valued in horticulture and grows on streets in many cities.

THE TRIASSIC PERIOD

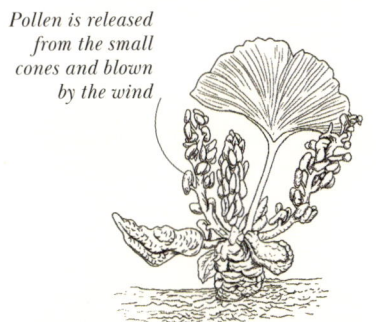

Pollen is released from the small cones and blown by the wind

Pollen cones

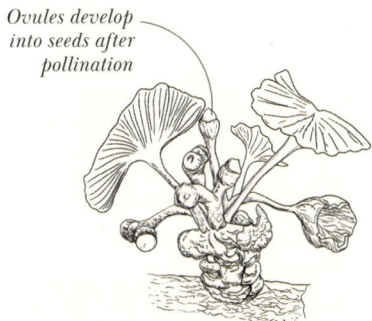

Ovules develop into seeds after pollination

Ovules

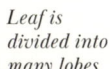

Leaf is divided into many lobes

Ginkgo fossil
Ginkgos were far more diverse in the past. In the Triassic, Ginkgoites produced leaves that were separated into many small lobes, rather than the simple fan seen in the modern species Gingko biloba.

RISE OF THE GINKGO

Ginkgos, which originated in the Permian, diversified globally during the Triassic, becoming a dominant seed-plant group. Today only a single species remains, but Triassic fossils reveal that the group was already diverse and widespread, with hardy forms adapted to the stressed landscapes that helped establish the first Mesozoic forests.

LYCOPHYTE RESURGENCE

The giant tree-lycophytes that had persisted since the Late Devonian finally went extinct at the end of the Permian. However, a new wave of much smaller members of the same group, the Isoetales, diversified into the harsh and disturbed environments of the Triassic. These species look much more akin to the only living relative of the group, *Isoetes*, which survives today in wetlands and lakes.

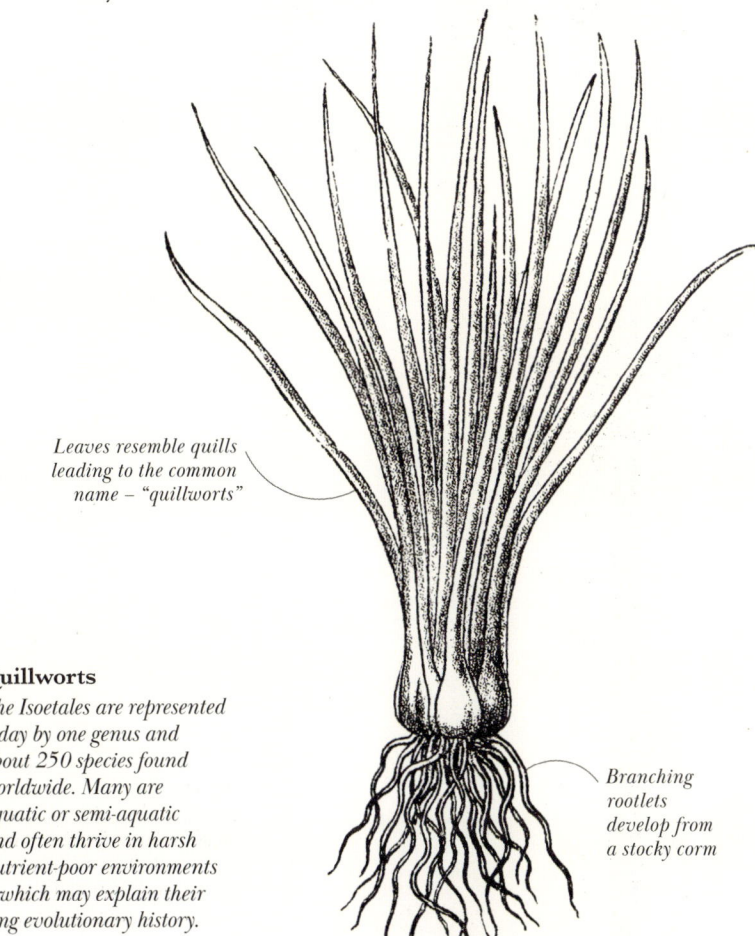

Leaves resemble quills leading to the common name – "quillworts"

Branching rootlets develop from a stocky corm

Quillworts
The Isoetales are represented today by one genus and about 250 species found worldwide. Many are aquatic or semi-aquatic and often thrive in harsh nutrient-poor environments – which may explain their long evolutionary history.

Hardy survivor
Pleuromeia was an Isoetales that spiked in abundance during the mass extinction at the beginning of the Triassic. The plant had an unbranched trunk and could reach a few metres in height – unlike the small, living Isoetales of today. It was also highly tolerant of environmental stress.

The reproductive cone is called a strobilus

Pleuromeia

Araucariaceae shed their lower branches, leaving a crown of branches at the top

PLANTS IN THE AGE OF DINOSAURS

The Jurassic Period spanned roughly 55 million years, from 200 to 145 million years ago. It was the heyday for seed plants and marked the peak of their ecological dominance before the later rise of flowering plants.

JURASSIC PARKLAND

The warm and humid Jurassic Period saw seed plants dominate, as tectonic forces started to break apart the supercontinent Pangea. Seed plants were globally distributed and adapted to distinct biomes. Lush conifer forests occurred in high latitudes and included groups that are familiar today, such as the relatives of monkey-puzzles, cypresses, and ginkgos. Ferns, especially the relatives of horsetails and the royal ferns, flourished in the understorey.

In more arid regions, plants produced scaly trunks and foliage with thick, waxy cuticles. The Bennettitales and the cycads both had these scaly trunks and looked superficially similar. However, they were distinct groups with very different reproductive structures. The Bennettitales developed a complex "flowerlike" structure, which was believed to have been pollinated by insects. This is just one example of the way that plants were coevolving with animals and fungi.

Familiar forms
Jurassic forests would have had some familiar elements to them, with royal ferns (Osmundales) in the understorey below towering conifers related to monkey-puzzle trees (Araucariaceae).

The fossil record
Many of the plants that flourished during the Jurassic were very similar to those of the Late Triassic. Sites in Yorkshire, northern England, have yielded particularly well-preserved fossils of flora from the Jurassic.

Williamsonia
Bennettitales

Ginkgo
Seed plant

THE JURASSIC PERIOD

Lush forests cover the higher latitudes where the climate is warm and wet

Pangea starts to break apart, splitting into the continents Gondwana and Laurasia

Extensive arid biomes form along the equatorial belt

The Early Jurassic world
Around 200–180 million years ago Pangea began to fragment and the Tethys Ocean started to widen in between.

Pseudoctenis
Cycad

Podozamites
Conifer

Dictyophyllum
Fern

Sphenopsid
Horsetail

CYCADS AND BENNETTITALES

The Jurassic saw widespread abundance of two groups of plants with scaly trunks and a very similar habit. The cycads and the Bennettitales are so similar in overall habit that it was long thought that they were closely related groups. However, careful investigation of the structure of their leaves, their stomata, internal wood, and their reproductive structures demonstrated clearly that they were very distinct and quite distantly related groups of seed plants. That two different groups developed a similar form suggests that stocky trees with scaly trunks may well have been particularly well-suited to life in the Jurassic.

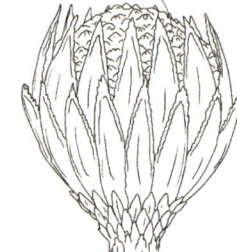

Domed structure composed of numerous stalked seeds

Williamsonia

Bract with pollen sacs

Weltrichia

Multiple bracts, each 3–5 cm (1–2 in) long, radiate out from a central cup

Weltrichia spectabilis

Telling them apart
One of the details that helps distinguish cycads and Bennettitales is their reproductive structures. As seen in this fossil, those of the Bennettitales are almost flowerlike in their complexity.

Male and female
Although they are named separately, Williamsonia and Weltrichia respectively represent the female and the male reproductive structures of the same group of plants.

DYNASTY OF THE ROYAL FERNS

The royal fern family (Osmundaceae), is an ancient group of ferns. Some of the living species are so similar in appearance to their extinct relatives that they have been considered to be living fossils. There are only about 20 species of Osmundaceae alive today, but as a group they have a spectacular fossil record, especially from the Permian, Triassic, and Jurassic. A remarkable Jurassic fossil was so well preserved that researchers were able to count the fossilized chromosomes. Remarkably, the number of chromosomes in the fossils was similar to the number in living relatives, suggesting there may have been comparatively limited changes in genome structure through time.

Fertile fronds are covered in spore-containing sporangia

Osmunda fern
There are three to six living genera in Osmundaceae, including Osmunda. It has fertile and sterile fronds or frond parts.

Veins in Dictyophyllum form a meshwork pattern

Dictyophyllum nilssonii
The royal ferns were not the only ferns to thrive in the Jurassic, other groups, such as the Dipteridaceae, were also widespread.

• THE JURASSIC PERIOD •

Cycad reproductive structures resemble large pine cones

Jurassic cycads were stocky trees that grew slowly but could live for hundreds of years

Stout, woody trunk with a scaly texture that formed from the bases of old leaves

Cycads
Based on their habit, the cycads look as though they have changed very little since the Jurassic, but it is now known that these plants are not living fossils. The living species are comparatively young, with other characteristics quite different from their ancestors.

ARAUCARIA AND FUNGI
Exceptionally preserved fossils from around the world, but especially from Argentina, provide insights into the amazing biology of Araucariaceae. Fossil evidence suggests that their roots formed specialized structures called root nodules. Fungi lived within these nodules and exchanged nutrients with the plants.

Jurassic plants had the same habit as the living monkey puzzle tree

Scaly tree stem

Monkey puzzle relatives
The Jurassic saw the rise of many groups of modern conifers that are familiar today, including the relatives of the monkey puzzle (Araucaria araucana).

Hyphae network (white) produced by the fungi

Mycorrhizae
Like the Jurassic Araucaria, the majority of living plant species form a beneficial symbiosis with fungi, called mycorrhizae. The fungi extract nutrients from the soil and exchange these with plants for carbon in the form of sugars produced during photosynthesis.

New forms evolve
Cretaceous deposits provide the first real evidence for flowering plants. Fossils of the extinct flowering plants Archaeanthus were discovered in western North America. The fossils are roughly 100 million years old and are most similar to modern tulip trees.

Spiral arrangement of flower parts characteristic of early angiosperms

FLOWERING PLANTS EMERGE

During the Cretaceous Period, roughly 145–66 million years ago, flowering plants, with their wonderfully diverse flowers, leaves, fruits, and seeds, evolved from the shadows to global dominance.

FLOWERS DOMINATE

The Cretaceous is one of the most important periods of plant evolution as it marks the rise to ecological dominance of the flowering plants (angiosperms).

The exact origins of angiosperms are uncertain and it remains unclear which extinct lineages of seed plants they are most closely related to. However, regardless of when they first emerged, it was during the Cretaceous Period that the angiosperms exploded in diversity. By the end of this period – when dinosaurs were still the dominant fauna and the continents had moved into a far more recognizable arrangement – the angiosperms were widespread and had split into many of the modern groups still alive today.

While angiosperms were rapidly diversifying, so too were the ferns, gymnosperms, and lycophytes. The hot-house conditions of the Cretaceous allowed lush ecosystems to thrive, even at high latitudes.

The fossil record
The more resilient lineages took advantage of new ecosystems to diversify, while others dwindled. Palms, such as Nypa, made their appearance in the Cretaceous.

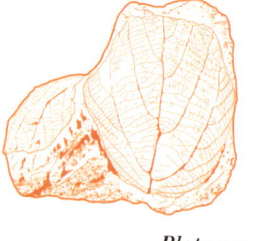

Platanus
Angiosperm

RETURNING TO WATER

Roughly 90 per cent of all living plant species are angiosperms and they dominate almost all biomes today. Part of the success of this group is due to its ability to pioneer new biomes and adapt to different modes of life. This is exemplified by the unprecedented number of angiosperms that made the move back into aquatic environments. Their aquatic diversity spans tiny free-floating plants, such as rootless duckweed at about 1 mm ($^{1}/_{32}$ in) in size, to the giant water lilies whose leaves can be 3 m (10 ft) in diameter. Angiosperms such as sea grasses and mangroves have even evolved into the marine saltwater environments. Fossils from the Cretaceous provide evidence for the presence of angiosperms in water, including relatives of modern lotus, water lilies, and extinct groups like *Archaefructus*.

Water lily

Water lilies grow in lakes and slow-moving water in temperate and tropical climates around the world. They are one of the earliest branches of the angiosperm family tree and have a fossil record extending back to the Cretaceous.

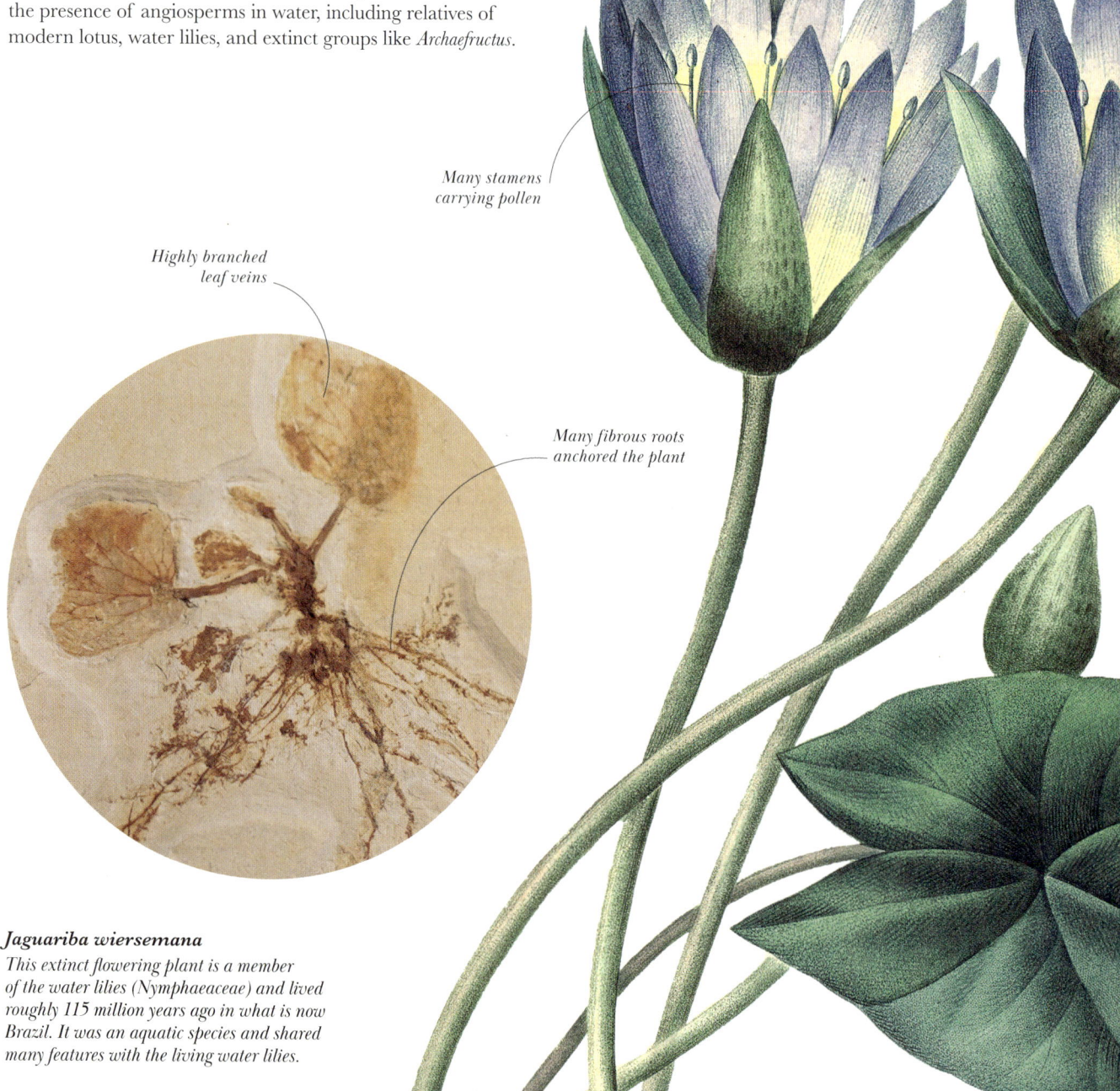

Many stamens carrying pollen

Highly branched leaf veins

Many fibrous roots anchored the plant

Jaguariba wiersemana

This extinct flowering plant is a member of the water lilies (Nymphaeaceae) and lived roughly 115 million years ago in what is now Brazil. It was an aquatic species and shared many features with the living water lilies.

• THE CRETACEOUS PERIOD •

Flower buds grow from a single stalk

Large flowers have petals in spiral pattern

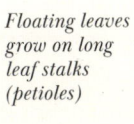

Floating leaves grow on long leaf stalks (petioles)

ADAPTING TO FIRE

Fossil evidence suggests that plants have been coevolving with fire for over 430 million years. Some plants therefore have adaptations to protect themselves against fire and many of these are seen in *Sequoia* (redwoods). *Sequoia* have thick bark that resists burning, they are able to resprout from dormant buds, and some species even require fire to complete part of their life cycle.

Sequoia dakotensis
The redwoods are an ancient group of conifers that was widespread in the Cretaceous. This fossilized cone is from an extinct ancestor.

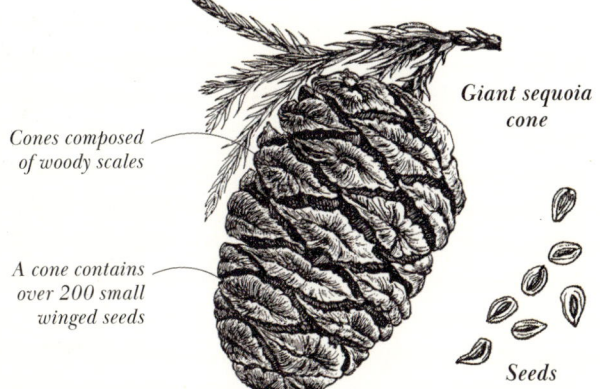

Giant sequoia cone

Cones composed of woody scales

A cone contains over 200 small winged seeds

Seeds

Needing heat
The cones of giant sequoia need the heat of a fire in order to open. Seeds are then released into the fertile ash-covered soil that is ideal for germination.

EXPLOITING NEW NICHES

As angiosperms rose to dominance, they created new biomes and niches for other plants to evolve into. Among those plants were spore plants, such as mosses, and spikemosses (a type of lycophyte). *Selaginella* was one of the most successful spikemosses in the Cretaceous, filling the shaded understorey niches.

Many tightly packed small leaves create a large frond for capturing light

Life in the shade
There are roughly 750 species of Selaginella today, and many are well adapted to shady environments. Some species have evolved blue iridescent leaves that are adapted for photosynthesis in low-light conditions.

Stems called rhizophores grow down and give rise to roots

Polar palms
Palms are evergreen and thrive in tropical climates. Early Paleogene fossil pollen suggests that the climate was suitable for palms to grow in polar regions.

Large, featherlike fronds of Nypa palms can be multiple metres in length

SURVIVING EXTINCTION

The Paleogene represents the first period of the most recent geological era, the Cenozoic. It lasted from 66–23 million years ago and the global climate was warm and wet, allowing tropical species to thrive.

RISE OF THE RAINFOREST

The Paleogene started with a bang when the asteroid that spelled the demise of the non-avian dinosaurs hit Earth. This catastrophic event caused a major upheaval to plant communities, causing local extinctions and the collapse of many ecosystems. The turmoil paved the way for ferns once again to briefly spike in abundance, before flowering plants firmly re-established themselves across the world. Flowering plants then dominated all major biomes and the warm Paleogene Period, with high atmospheric CO_2, supported the diversification of a lush, tropical flora across the world.

The Paleogene also marked the origin of modern rainforests. These forests included a closed canopy where the crowns of trees blocked light, forming a darker, more humid understorey where shade-tolerant species thrived.

Globally, the movement of continents led to the growth of mountains such as the Himalayas, and monsoon conditions were established over parts of Asia.

The fossil record
Fossils from the Paleogene reveal the characteristic leaves, propagules, and pollen of flowering plants (angiosperms). Fossilized propagules of mangroves preserve evidence of coastal vegetation, and this period saw many angiosperms diversify to colonize more aquatic habitats.

Dipterocarp
Tropical tree

Legume roots
The symbiotic bacteria that fix nitrogen are very sensitive to high levels of oxygen. Plants therefore use a pink iron-containing molecule to reduce oxygen levels.

Large root nodules create a stable, low-oxygen environment for bacteria

Five petals are arranged in a distinct fashion that is ideal for insect pollination

Tendrils are specialized leaves that help the plant climb

Pea plant
The legumes, the Fabaceae family, contain almost 23,000 species, including peas, lentils, soya beans, and peanuts. They can usually be recognized by their seed pods and distinct flowers. The majority of species form a symbiosis with nitrogen-fixing bacteria.

NITROGEN NODULES

Plants use nitrogen for a variety of processes, including to create chlorophyll for photosynthesis. Although nitrogen makes up 78 per cent of the atmosphere, it exists in a form that cannot readily be absorbed by plants. Plants instead have to invest lots of energy producing extensive roots to forage in the soil for nitrogen that has been broken down by microorganisms. However, in the Paleogene a group of plants called the legume family evolved a new strategy to access nitrogen. They formed a symbiosis with a type of bacteria called *Rhizobia* that is capable of fixing nitrogen from the atmosphere into a usable form for plants. *Rhizobia* exchange this nitrogen for sugars from the plant and the plant's root nodules provide a stable environment for the bacteria to live.

AZOLLA EVENT

Flowering plants were not the only group of plants to evolve back into the water (see p.46) – ferns made this transition too. The return to water of one group of ferns, called *Azolla*, may have had a global significance. About 50 million years ago, Earth's climate was much warmer than today, largely due to greenhouse gases in the atmosphere. Around this time, *Azolla* achieved exceptionally large blooms. It formed vast floating mats close to the North Pole, which was ice-free at the time. This bloom was so large that one theory suggests that it drew down great amounts of carbon dioxide (CO_2) from the atmosphere and may have contributed to a drop in global temperatures.

Overlapping leaves form dense mats

Aquatic fern
Azolla may not look like a typical fern but it is a free-floating fern found in freshwater habitats.

PUTTING RAIN IN THE RAINFOREST

One of the remarkable innovations of angiosperms was a change in their capacity to transport water. Angiosperm leaves have much higher densities of veins for water transport and many more small pores (stomata) on their surfaces. This means that, compared to their ancestors, they could transport and release a lot more water through transpiration. Occurring on a global scale, it is thought that angiosperms helped to create rainforests.

Chew marks have formed holes

Ancient interaction
Fossil angiosperm leaves provide direct evidence of high vein densities in the past. This one also shows evidence of damage from an animal.

The base of the flower has five fused sepals, which form a green, cuplike structure

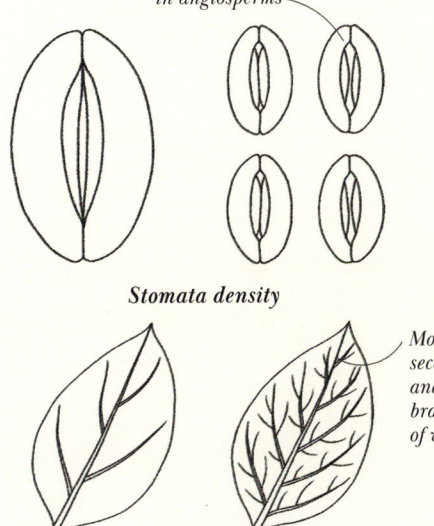

Stomata are smaller and occur at a higher density in angiosperms

Stomata density

More secondary and tertiary branching of veins

Leaf-vein density

Stomatal density
The origin of angiosperms led to far greater leaf-vein densities and a tenfold increase in stomatal density.

Leaves have a drip tip to shed heavy rainfall

Leaves are arranged to maximize the exposure to sunlight

Rainforest foliage
Plants in warm tropical rainforests are typically broad-leaved evergreens with adaptations for heavy rain.

Taking over
The grass family (Poaceae) is a large group of flowering plants that consist of nearly 12,000 species. Grasses now have a global distribution and dominate many biomes, including prairies, savannas, and steppes.

Many small, wind-pollinated flowers called florets are clustered together in a spikelet

THE EXPANSION OF GRASSLANDS

The Neogene, roughly 23–2.6 million years ago, was a period of global cooling, drying, and mountain-building. These processes helped to separate plant groups into distinct biomes and the period also saw the rise of grasses.

A WORLD OF NEW OPPORTUNITIES

As Earth's continents moved to their modern positions during the Neogene, mountains such as the Himalayas and Andes continued to form. Over time, the climate switched from being warm and wet to becoming cooler and drier. A dry world with dramatic mountainous regions gave plants many opportunities to specialize and new biomes, such as savannas, dry forests, deserts, cloud forests, and upland plateaus became established. Distinct plant groups came to define these new biomes and supported complex ecosystems.

The greatest botanical change of the Neogene was the evolution of grasses. Forests retreated, giving rise in their place to huge areas of grasslands. Grasses saw an explosion in species richness and were armed with adaptations to help them thrive, including a new, highly efficient variation of photosynthesis. While grasslands expanded, Earth continued to cool and towards the end of the Neogene, the icecaps forming at the poles were the first sign of an approaching Ice Age.

The fossil record
Plants like Arbutus grew in Mediterranean-type biomes, with wet winters and dry summers. Known from fossilized pollen, the grassland shrub Ephedra thrived in even drier environments.

Picea
Conifer

Fagus
Angiosperm

THE AGE OF GRASSLANDS

As grasses diversified, they evolved a new, more efficient mode of photosynthesis, referred to as C4 photosynthesis. C4 plants are uniquely adapted for environments that are hot, dry, and have high light intensities, such as savannas. They became globally widespread in the last 10 million years.

Brachypodium sp. (C3)

Themeda sp. (C4)

Invisible difference
C3 and C4 grasses can only be distinguished visually by studying their leaves under the microscope.

Hairlike awns help the seeds disperse

Individual flowers are called florets

Inconspicuous flowers
Grass flowers are small and lack showy petals, in part because they are wind-pollinated and do not have to attract animal pollinators.

Fruit has a hard shell and resembles a coconut

Whole fruit

Shell contains a large number of wedge-shaped, edible seeds

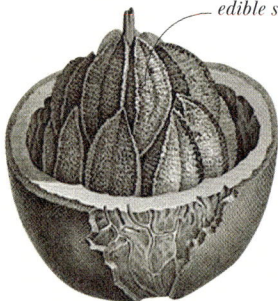

Inside fruit

Brazil nuts
The term Brazil nut can refer to both the edible seeds and the trees that produce them (*Bertholletia excelsa*). The trees are indigenous to the Amazon rainforest and live for hundreds of years.

EVOLUTION OF NORTHERN PEATLANDS

As Earth cooled during the Neogene, there was an enormous expansion of cool, wet peatlands in the Northern Hemisphere. This expansion was facilitated by the diversification of a group called the bog mosses, *Sphagnum*. There are over 300 species of *Sphagnum* today and almost all are thought to have evolved during the Neogene. *Sphagnum* has an amazing capacity to store water, and helps maintain peatlands that lock up enormous reserves of carbon.

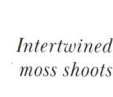

Intertwined moss shoots

Peat plant
As Sphagnum dies, the waterlogged conditions it lives in means that it decomposes slowly and not completely. Over time, it is compressed into carbon-rich peat.

THE RAINFOREST IS BORN

The Neogene saw the gradual change of rainforests to their modern form and extent. Fossilized amber from China indicates that during the early stages of the Neogene there was a much greater expanse of Asian tropical forests than there is today. These forests gradually retreated to the tropics as Earth cooled. While it originated in the Paleogene, the Amazon rainforest took on its modern form during the Neogene, when tectonic activity resulted in the creation of the eastward-flowing Amazon River; previously the river flowed to the west. When its direction of flow was reversed, it started draining into the Atlantic Ocean, turning the vast wetland biome that was there at the time into a modern rainforest one, with characteristic plants such as the Brazil nut.

THE NEOGENE PERIOD

New leaves are initially rolled up and protected by a reddish or pinkish sheath

Glossy, leathery leaves have smooth margins

Ficus elastica
The rubber plant is a tropical evergreen tree that grows in wet tropical biomes and is indigenous to South and Southeast Asia. It can reach more than 50 m (165 ft) in height, supported by extensive aerial and buttressing roots. Distinct fossil leaves demonstrate that Ficus species grew in Southeast Asia during the Neogene.

Large leaves grow up to 30 cm (12 in) in length

At home on the tundra

Cottongrass (Eriophorum) is a flowering plant and part of the sedge family. It is found in bogs in the Northern Hemisphere and the Arctic tundra, but was more widespread in the Early Quaternary.

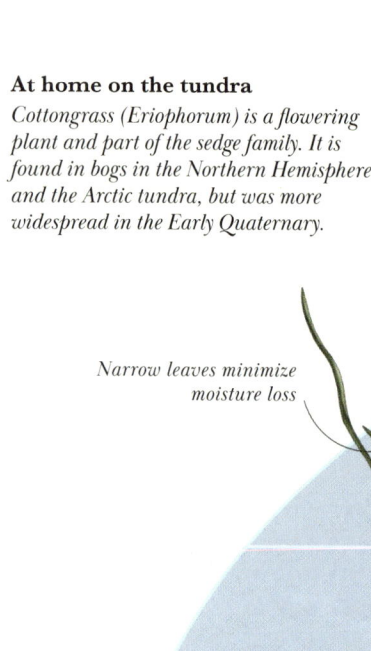

Narrow leaves minimize moisture loss

NORTH AMERICA

Much of North America and northern Europe were covered by ice

PACIFIC OCEAN

ATLANTIC

SOUTH AMERICA

The dry climate adds to the expansion of desert biomes

SURVIVING THE ICE AGE

The Quaternary Period spans the last 2.6 million years of Earth history through to the present day. This was an Ice Age, where Earth oscillated between cold glacial and warmer interglacial periods.

LAND OF MAMMOTHS

Rather than one giant freeze, the Quaternary world experienced a cycle of cold and warmer periods, with ice sheets and different biomes expanding and contracting in response to the changing climate. This cool world supported the evolution of the mammoth steppe, also known as the steppe tundra. This new biome stretched across vast swathes of Eurasia and North America, forming one of the largest continuous ecosystems in Earth's history. It was a highly productive environment, dominated by sedges, grasses, forbs, and other herbaceous species, and was home to large herbivores such as the woolly mammoth. During the warmer interglacial periods, forests also spread into the steppe.

Warming towards the end of the Quaternary Period eventually led to the disappearance of the mammoth steppe and the retreat of the ice sheets to their modern levels. Away from the ice, our human ancestors were spreading and domesticating crop species, which would eventually lead to modern agricultural biomes.

THE QUATERNARY PERIOD

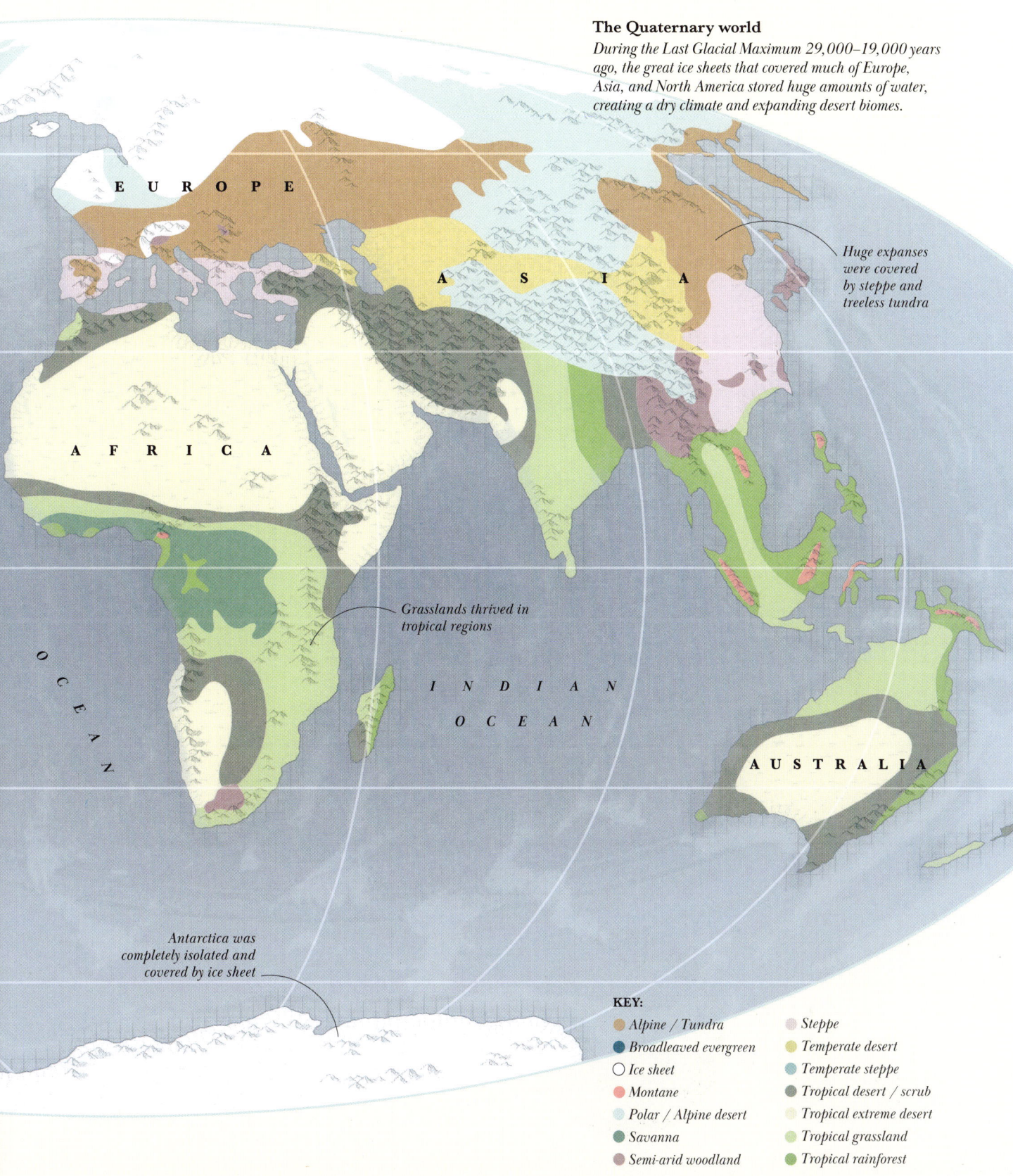

The Quaternary world
During the Last Glacial Maximum 29,000–19,000 years ago, the great ice sheets that covered much of Europe, Asia, and North America stored huge amounts of water, creating a dry climate and expanding desert biomes.

Huge expanses were covered by steppe and treeless tundra

Grasslands thrived in tropical regions

Antarctica was completely isolated and covered by ice sheet

KEY:
- Alpine / Tundra
- Broadleaved evergreen
- Ice sheet
- Montane
- Polar / Alpine desert
- Savanna
- Semi-arid woodland
- Steppe
- Temperate desert
- Temperate steppe
- Tropical desert / scrub
- Tropical extreme desert
- Tropical grassland
- Tropical rainforest

THE MAMMOTH STEPPE

A steppe biome is a large, dry, grassy plain characterized by a semi-arid climate. The plants that thrive here are short, drought-resistant grasses and shrubs with deep roots. Today, steppes can be found across Eurasia, the Great Plains in North America, and Patagonia in South America. In the glacial conditions of the Quaternary, a novel type of steppe evolved. The mammoth steppe was a vast ecosystem whose plant species were adapted for life above a layer of permanently frozen soil called permafrost. The permafrost ensured that nutrients and moisture were kept near the surface, creating a productive ecosystem, but the layer of frozen soil also prevented trees with deep root systems from becoming established. At its peak, the mammoth steppe extended across northern Europe, Asia, and North America.

The wind-dispersed seed heads resemble tufts of cotton

Tundra species
Cottongrass (Eriophorum) are sedges that can live in wet, acidic soil and are commonly found in Arctic tundra biomes.

High plateau
The cold, high-altitude steppe environment is characterized by vast, open grasslands, like the ones on the Siberian plateau.

SURVIVORS FROM THE PAST

More than 99 per cent of all species that have ever lived are now extinct and only a tiny percentage of these are preserved as fossils. With species gradually going extinct through time, we are losing living connections to key groups of plants that thrived in the past. In some rare cases there is the chance to catch a glimpse of the very tail end of these lineages before they are lost to extinction. For example, the Wollemi pine is the sole living relative of its genus and was only discovered in 1994. The dawn redwood was identified first from fossils before a tiny living population was described in 1948.

Last chance to see?
The Wollemi pine (Wollemia nobilis) was discovered from a small population of mature trees in New South Wales, Australia. Amborella trichopoda is the only surviving member of the Amborellaceae – the sister branch to all other living flowering plants.

Wollemi pine

Wollemi cones are large, like those of monkey puzzle trees

The outer parts of the small female flower are not differentiated into distinct petals and sepals

Amborella

THE DOMESTICATION OF CROPS

It is estimated that about half of the world's habitable land is used for agriculture today. This is a giant increase from an estimated 4 per cent 1,000 years ago. The plants that are cultivated today have been changed from their wild ancestors to make them ideal agricultural crops through the process of domestication. The domestication of wheat, rice, and maize occurred gradually over the last 10,000 years, as people selected plants to propagate that had desirable traits, such as reduced branching and seeds that were not shed when mature. Domestication underpinned the rise of human civilization.

Domesticated maize
During domestication, humans selected for features that would make maize ideal as a food crop. This led to a reduction in stem branching so that the plant could put its energy into growing more and significantly larger kernels that had no hard case around them.

Bushy, highly branched growth form

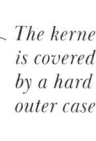

The kernel is covered by a hard outer case

Terminal ears are small and contain only a few kernels

Teosinte maize
Maize was domesticated from teosinte approximately 9,000 years ago, in the Balsas River valley in Mexico. The wild plant was originally grown as a food crop and possibly used like sugar cane for its sweet stems.

Section 2

MODERN BIOMES

MODERN BIOMES

The geographical areas and specific conditions in which plants live can be defined as biomes – broad, often very familiar environments such as forests and deserts. There have been many attempts to classify these biomes in greater detail, and the world of plants is better understood by looking at more nuanced sub-biomes within each type of habitat.

The concept of a biome is useful for exploring the diversity and distribution of life on Earth. Many of these categories are instantly recognizable; it is easy to distinguish a forest from a desert or a swamp by the water, soil, climate, and the types of plants and other organisms that live there. A forest, for example, has a fairly dense cover of trees as well as shrubs and herbaceous plants, and a profusion of associated animals and fungi. However, within that broad forest category, the climate, seasonality, altitude, and the type of soil all add nuance. Boreal forests are vast, cold, and dominated by conifers, while temperate forests are strongly influenced by seasonality and feature different combinations of plants, animals, and fungi, all of which – together and separately – affect how life operates in those environments.

PATTERNS OF CHANGE
Modern biomes are structured by the plants that live in them, so it is natural to describe a biome according to the plant life that it hosts. The vast steppe and prairie grasslands of the temperate zone are relatively dry, inland areas that are dominated by well adapted grasses – in tropical grasslands, a sparse patchwork of trees and shrubs often appears, forming savanna.

In all grassland biomes, the interaction with grazing mammals or fire is crucial to maintain the grassy ecosystem that makes it hard for woody plants to establish. Where fire or grazing are reduced, grasslands may shift over time to become scrubbier and ultimately dominated by trees – effectively they transition to a forest biome, where competition for light and space become the primary challenges.

Such transitions result in overlapping biomes: the Mediterranean biome, for instance, is a mixture of forest, shrublands, and scrub scattered over at least five disconnected areas, and may be considered part of the forest biome and the grasslands and shrublands biome. Similarly, vast reaches of Siberia and the Canadian interior are an intricate mosaic of steppe grassland and cold shrublands, wetlands, and boreal forests.

Patchwork planet
Although they intersect, biomes can be divided into simple categories such as those defined by the Worldwide Fund for Nature (WWF). They include forests, grasslands and shrublands, deserts, mountains, and wetlands. The newest biome – human environments – overlaps with them all.

MODERN BIOMES

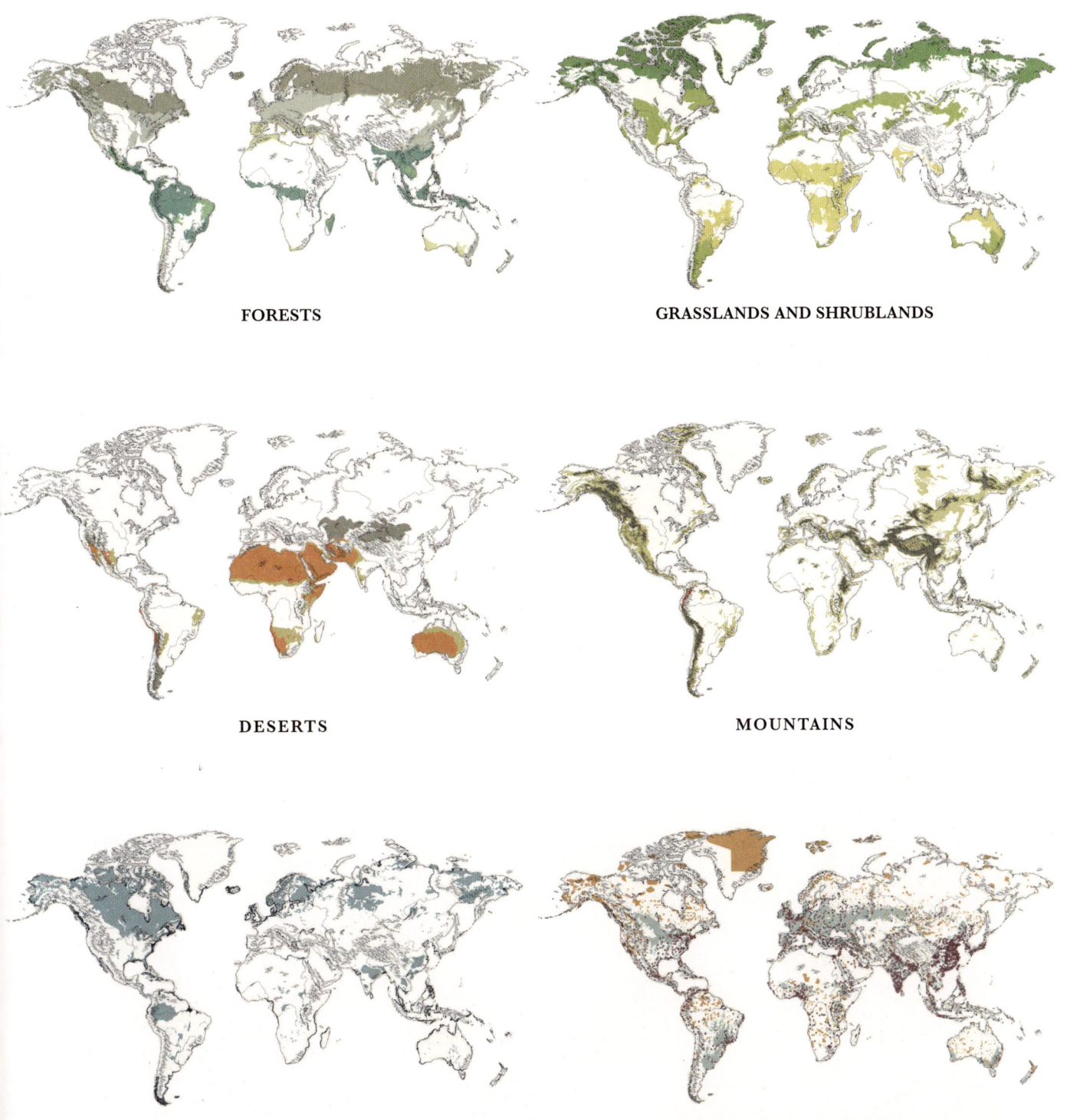

Every biome presents an array of niches and specific conditions to which plants must adapt in order to survive and reproduce. Plants that cannot adapt, or move to a more suitable place, face extinction. Success in these niches is often a matter of chance mutations, combined with being in the right place at the right time to establish, as well as the ability to spread.

COPING STRATEGIES

Plants that adapt to the conditions in a biome can avoid competition and exploit resources that would not otherwise be available. Desert plants, for example, share a suite of adaptations that allows them to cope with fierce temperatures and low rainfall – succulence to store water, thickened, waxy cuticles, and spines for protection. A diverse range of plants often respond to challenging conditions in similar ways, so the same solutions may appear in completely unrelated groups.

In flowering plants, carnivory has evolved – usually as a response to poor nutrient conditions – at least ten times. Multiple species from different families – and on different continents – have modified their leaves to catch insects on sticky surfaces, in elaborate pitchers, or inside speedy snap-traps. The mechanisms and leaf parts involved vary in subtle ways, but the use of the same structure to do a similar job occurs repeatedly.

SPECIALIZATION AND DIVERSITY

Sometimes a plant has adapted so closely to a niche that a change can spell disaster for the species. This is particularly the case where a plant relies on a specific pollinator for successful reproduction: if the pollinator is lost, so is the plant. These kinds of co-dependencies have a way of driving diversification, because the plant and its pollinator must keep up with each other if they are to survive. Non-living (abiotic) factors may define the baseline of a biome, but more often the true driver of diversity is the interaction between living species – the biotic factors. The interplay between abiotic factors and biotic interactions present numerous challenges and opportunities for every species to adapt, adding layers of complexity – and diversity – to every biome.

fig. 1 Opuntia ficus-indica; fig. 2 Lathyrus oleraceus; fig. 3 Dionaea muscipula; fig. 4 Lithops hookeri; fig. 5 Euphorbia pulcherrima; fig. 6 Kalanchoe pinnata; fig. 7 Allium schoenoprasum; fig. 8 Salvinia natans

Reduced leaves become hard, sharp spines

Coiling tendrils are modified leaflets that pull plant up supports

Hinged traps are formed from leaf blades

fig. 1   fig. 2   fig. 3

DEFENSIVE LEAVES   **LEAVES THAT CLIMB**   **CARNIVOROUS LEAVES**

• MODERN BIOMES •

Versatile leaves

Leaves are the Swiss Army knives of the plant world. They are primarily designed to perform photosynthesis, but sometimes plants deploy them in ingenious ways to cope with specific conditions, or compensate for features such as tiny flowers that pollinators might overlook.

Translucent surface allows light to reach chlorophyll held in cool underground parts of the thick truncated leaves

fig. 4

LEAVES WITH WINDOWS

Highly visible leafy bracts attract pollinators to the tiny flowers

fig. 5

COLOURFUL BRACTS

Plantlets (pups) on leaf margins fall and root to make new plants

Fleshy leaves store sugar and water in underground bulb during dormancy

This aquatic plant's highly dissected leaf absorbs nutrients from water

fig. 6

LEAVES THAT REPRODUCE

fig. 7

LEAVES THAT STORE FOOD

fig. 8

ROOTLIKE LEAVES

Chapter 1

FORESTS

Soil pH

Most forests soils are slightly acidic due to decomposing organic material and the composition of the underlying rocks. Slightly acidic soils make soil nutrients more soluable, and easier for forest plants to absorb through their roots.

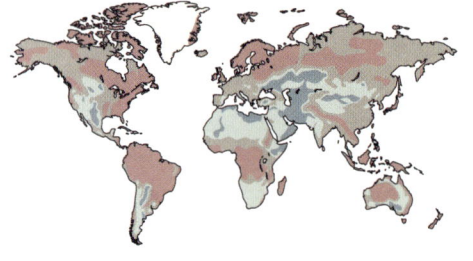

KEY
- pH 8 (alkaline)
- pH 7 (neutral)
- pH 6 (acidic)
- pH 5 (acidic)

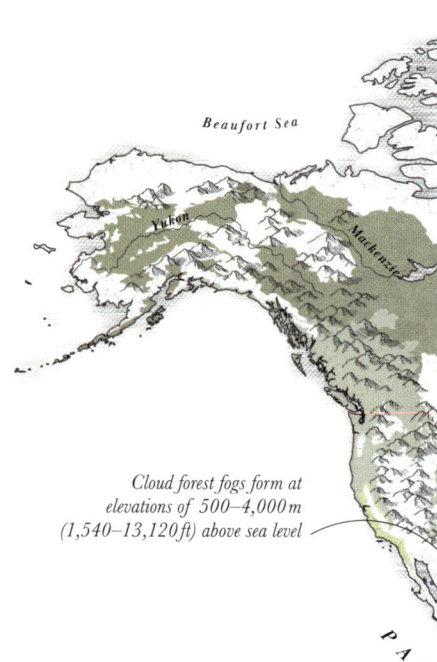

Cloud forest fogs form at elevations of 500–4,000 m (1,540–13,120 ft) above sea level

The Mediterranean forest in Chile was formed in isolation by the rise of the Andes

FORESTS

Covering almost a third of Earth's total land surface, forests span extreme climate zones, from wet, hot tropics to dry, cold boreal regions. They are characterized by dense communities of trees, but many other plants thrive in their shade.

A DIVERSE BIOME

Temperature, soil fertility, elevation, and precipitation levels determine both the distribution of forests and the characteristics of plants within them. Although forest ecosystems are dominated by trees, they are incredibly biodiverse, hosting communities of many different plant and fungi species that both support and compete with each other.

Rain and shine

Tropical and subtropical forests experience little annual temperature variation. Other forests are at their hottest in summer, which generally corresponds to periods of high or low precipitation. In these forests, plants have adapted the timing of their main growing season to periods when the balance between temperature and precipitation is ideal.

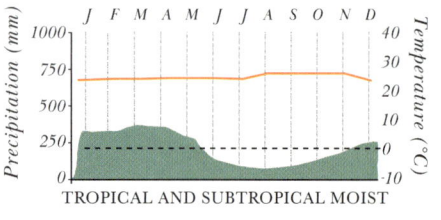
TROPICAL AND SUBTROPICAL MOIST

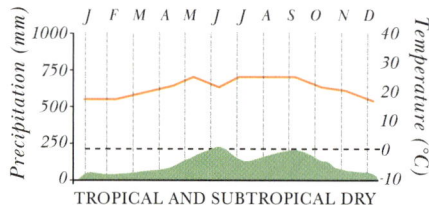
TROPICAL AND SUBTROPICAL DRY

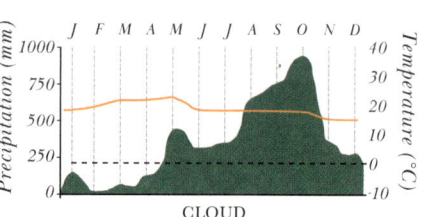
CLOUD

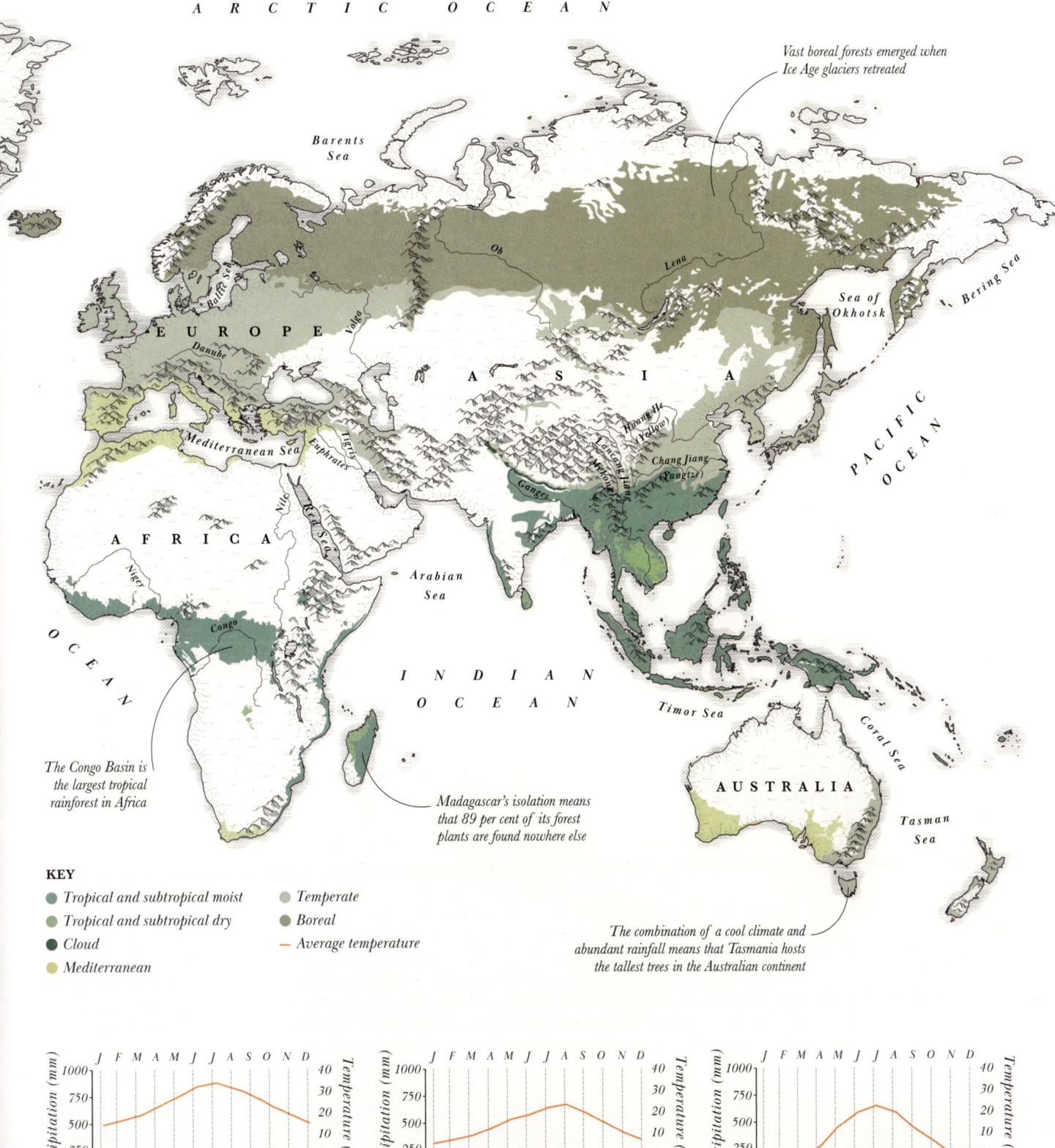

Seeking sunlight

Forest plants use a range of strategies to capture sunlight. Some plants angle their leaves precisely to reduce self-shading; others have green stems that help to carry out photosynthesis; and many grow darker green shade leaves packed with chlorophyll to absorb more light on the dimly lit forest floor.

Leaf position

fig. 1

- Some leaves grow in pairs on opposite sides of the stem
- Pairs of leaves set at different angles reduce shading from foliage higher up the stem

Photosynthetic stem

fig. 2

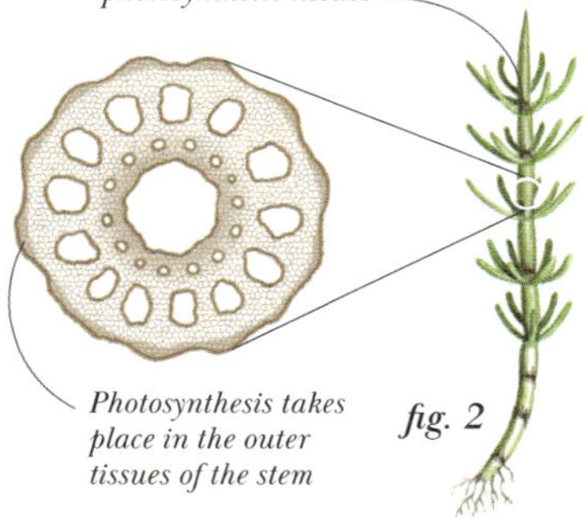

- Horsetail stems have photosynthetic tissues
- Photosynthesis takes place in the outer tissues of the stem

Darker leaves

fig. 3

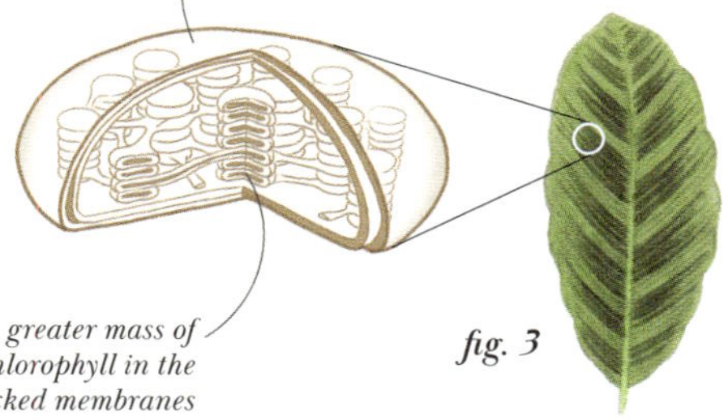

- Shade leaves contain chlorophyll-carrying chloroplasts that move around the cells to areas in light
- A greater mass of chlorophyll in the stacked membranes aids photosynthesis

Elongated stem

fig. 4

- In full light, the stem is short and the leaves crowd together
- In shade, the stem extends and leaves spread in search of light

Giant leaf surface

fig. 5

- Large leaves spread horizontally to capture more light

Thriving in shade

Some forest plants are adapted to grow in deep shade, especially in the understorey and on the forest floor. They may develop oversized leaves to catch scattered light, or elongated stems that help them reach pockets of light from gaps in the canopy.

- Broad, dark leaves help the cacao tree to absorb light in the shady understorey

EMERGENT LAYER
- The emergent layer is home to huge numbers of epiphytes

CANOPY LAYER
- Lianas scale the tallest trees, scrambling through the canopy to reach and capture direct sunlight at the very top of the trees

UNDERSTOREY

FORESTS

SHRUB LAYER

FOREST FLOOR

Rainforest layers
Emergent trees tower above the canopy to reach full sunlight, while the understorey, shrub layer, and forest floor lie in deep shade. Some plants grow upwards to reach the light; others thrive in the lower, darker levels.

Space for roots
Roots anchor the plant and absorb nutrients. As forest plants compete for space and light, their roots also adapt. Epiphytic and aerial roots draw in moisture from the air, while shallow roots knit together for stability.

Roots are shallow to share space with other trees nearby

Aerial roots grow down to the ground and support branches as they spread in search of more light

Buttress roots spread wide at the base of tall trees, helping to stabilize them in shallow or wet soils

Epiphytes grow on the branches of other plants

Roots anchor the plant by clutching or wrapping around tree branches

Roots spread horizontally and intertwine with others

Easy access to nutrients allows the plant to grow new shoots and leaves

Aerial roots have less competition for space and nutrients

fig. 7
fig. 8
fig. 6
fig. 9
fig. 10
fig. 11

Epiphytic roots
Knitted roots
Aerial roots

fig. 1 *Catharanthus roseus*; fig. 2 *Equisetum sylvaticum*; fig. 3 *Calathea sp.*; fig. 4 *Nicotiana tabacum*; fig. 5 *Philodendron giganteum*; fig. 6 *Ceiba pentandra*; fig. 7 *Ficus benghalensis*; fig. 8 *Theobroma cacao*; fig. 9 *Tillandsia sp.*; fig. 10 *Sequoia sempervirens*; fig. 11 *Ficus benghalensis*

ADAPTATIONS
LIGHT AND SPACE

Competition for light and space shapes the structure of forests and drives many of the adaptations in the plants that live in these crowded environments. Some, such as the kapok tree (*Ceiba pentandra*) and banyan (*Ficus benghalensis*), grow tall or develop broad canopies to capture sunlight. Others climb or live as epiphytes – plants that grow on the surface of other plants without harming them – using their hosts for support as they reach higher into the light and space. Meanwhile, in the dimmer understorey, shade-tolerant species such as the cacao tree (*Theobroma cacao*) make the most of limited light, producing fruit far beneath the forest giants.

Storage organs

Plants living in the woodland understorey have an array of storage organs that provide a food reserve for seasons when nutrients are scarce. This stored food allows the plant to flower and flush into leaf before the trees above do the same and block light to the forest floor.

fig. 1 — Rhizome — *Slender underground stem helps the plant grow quickly when it ends its winter dormancy*

fig. 2 — Root tuber — *Small root tubers can detach and grow new plants*

fig. 3 — Bulb — *Bulbs are specialized underground storage organs formed from modified leaves*

fig. 4 — Tuber — *Swollen tubers store energy as starch to kickstart plant growth in spring*

fig. 5 — Bulbil — *Bulbils – semi-mature plants – drop and quickly develop into new individual plants*

fig. 1 Anemone nemorosa; fig. 2 Ranunculus ficaria; fig. 3 Hyacinthoides non-scripta; fig. 4 Cyclamen sp.; fig. 5 Allium paradoxum; fig. 6 Quercus robur; fig. 7 Ilex aquifolium; fig. 8 Pseudotsuga menziesii; fig. 9 Fagus sylvatica; fig. 10 Acer rubrum

ADAPTATIONS

SEASONS

Although the cloud-forest habitats of the tropics tend to have a relatively stable climate, most other forest biomes are seasonal. The seasons vary by latitude and hugely influence the forest type – from the seasonally dry forests of the subtropics and tropics to the four-season forests of temperate zones. Trees and the plants under them have adapted to cope. In temperate forests, deciduous trees drop their leaves for winter when sunlight is brief and weak, while herbaceous plants retreat underground – then re-emerge before the leaf canopy regenerates.

• FORESTS •

Ready for winter

Leaves use a lot of energy, and as days shorten and sunlight becomes weaker, deciduous trees shed their leaves in order to conserve resources. Evergreen plants retain their leaves to capture light in darker days for photosynthesis.

Leaves drop when they are no longer needed for photosynthesis

fig. 6

Deciduous leaves

Tough, spiky leaves can survive winter

fig. 7

Evergreen leaves

Yellow carotenoid compounds are always present in leaves but become visible in autumn as green chlorophyll breaks down

Carotenoids in leaves reflect yellow and orange light

Tough evergreen needles reduce water loss in dry winter air

Abscission zone

Leaf is severed from plant as the abscission zone softens and breaks

Corky layer develops within abscission zone to seal plant wound

Slow-maturing cones hold seeds for dispersal in seasonal high winds

fig. 8

Leaf drop

Autumn colours

Evergreen needles

Green chlorophyll breaks down faster than yellow carotenoids

Sugars encourage cells to produce anthocyanin, which creates vivid red colours

fig. 9

Colour change

Chemical changes

Cool, dry autumn weather triggers hormonal processes in deciduous trees and shrubs that cause leaves to die. Leaf colours change as chlorophyll and carotenoids break down, and starches turn to sugars.

fig. 10

Turning red

Leathery leaves have a glossy dark green upperside

1 /

1/ Healing latex
Palaquium gutta
This tropical Asian evergreen thrives in moist climes in well-drained, fertile soil. It produces a sticky, milky fluid, or latex, which oozes from its bark when damaged, and thickens upon exposure to air. In this way, it seals wounds inflicted by insects or herbivores, and prevents further damage by repelling or trapping its attackers. The hardened latex, known as gutta-percha, is a natural form of rubber that has been in use since the 19th century.

PLANTS OF
TROPICAL AND SUBTROPICAL MOIST FORESTS

Tropical moist forests form a broad green belt around the equator, covering Central and South America, central Africa, Southeast Asia, and many Pacific islands. They are characterized by year-round warmth and heavy rainfall, which together create biodiversity hotspots. These forests support complex ecological structures, with towering trees forming dense canopies that capture sunlight, beneath which dimly lit understories support ferns, mosses, and shade-loving plants. Constant moisture fuels rapid growth, but also brings challenges: plants endure endless rain, relentless herbivory, and fierce competition for light.

Leaflets have a silvery white underside

2/

2/ Anchoring barbs
Calamus discolor
Found in the rainforests of tropical Asia, most rattans are climbing palms with long, whiplike stems that reach towards the forest canopy. Their leaf sheaths and petioles bear backward-pointing spines that snag anything brushing past – including fur, feathers, and skin. These spines are both a defence mechanism and allow the plant to grip surrounding trees as it clambers through. Rattans have strong, flexible stems that bend without breaking to avoid wind damage.

3 /   5 /

4 /   6a /   8 /

7 /

• TROPICAL AND SUBTROPICAL MOIST FORESTS •

3/ Fruits on stilts
Pandanus tectorius
Unlike woody trees, monocots such as pandanus do not support themselves with thick trunks. Instead, this plant props itself up with clusters of adventitious stilt roots that spread from its stem, giving it a mangrovelike appearance. This keeps it upright in the loose, sandy, or unstable coastal soils of Southeast Asia, Australia, and the Pacific Islands, where its starchy fruits attract a range of bird and animal seed dispersers.

4/ Top banana
Musa ingens
High in the Arfak Mountains of New Guinea and the forests of Indonesia, this plant towers like a tree, sometimes exceeding heights of 15 m (50 ft) tall, making it the world's tallest banana species. However, its lofty trunk is not made of wood, but consists of a column of overlapping leaf bases. This stem produces an inflorescence that can carry more than 300 fruits, with each banana up to 30 cm (12 in) in length.

5/ Trick and treat
Couroupita guianensis
Found in the rainforests of Central and South America, the cannonball tree – named for its large round fruit – sprouts large numbers of bright orange-red flowers directly from its trunk and main branches. While they do not produce nectar, these highly fragrant blooms invite carpenter bees and bats to feast on large amounts of sterile "false pollen", while dusting them with real pollen, to ensure successful pollination.

6/ Winged wonders
Shorea siamensis
A member of the dipterocarp family, characterized by their tall, straight trunks and winged fruits, this tree grows in lowland forests in Borneo. Its fruits have distinctive, papery wings, the largest reaching around 12 cm (5 in) long. These aerodynamic aids help the fruits to catch gusts of wind and glide away from the parent tree.

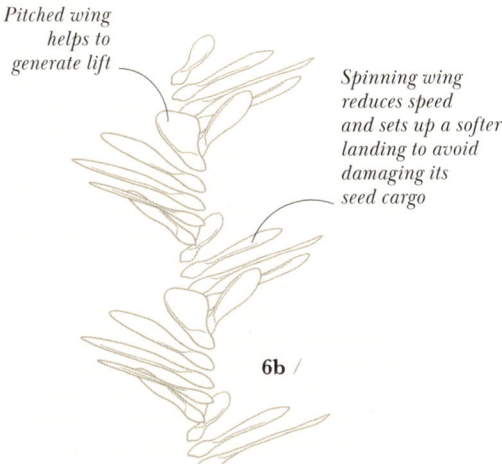

Pitched wing helps to generate lift

Spinning wing reduces speed and sets up a softer landing to avoid damaging its seed cargo

6b/

Going the distance
Dipterocarps' winged fruits, resemble tiny propellers, that whirl through the air – small ones drift nearby, but the largest fruits can ride a storm for several metres.

7/ Hole and hearty
Monstera deliciosa
This popular houseplant is native to tropical forests in Central America, where it climbs tree trunks using its aerial roots. Its huge, perforated leaves help it thrive in wet, low light conditions. The holes allow the each leaf to spread further, to help them capture the limited sunlight in the understorey. The leaves also act as drains, ensuring rainwater reaches the plant's roots.

8/ Dangerous flavour
Piper auritum
This member of the pepper family grows along shady riverbanks and in lowland tropical forests in Central and South America. Its leaves and young stems contain a distinctive aromatic oil that is rich in safrole, a compound with a spicy, anise-like flavour. This oil acts as a chemical deterrent, discouraging herbivores and insects from feeding on the plant – for some insects ingesting it is fatal.

9/ Toxin trade-off
Chondrodendron tomentosum
In the tropical forests of South America, browsing parts of this plant can stop animals in their tracks. Its bark and stems contain higher concentrations of curare, an alkaloid that blocks muscle control upon entering the bloodstream, causing paralysis and even death. However, some parts provide a safer meal – the plant relies on seed dispersers such as birds and monkeys to eat its berrylike fruit.

Purplish black fleshy fruit resemble large grapes

9/

Female cones are broadly ovoid and borne at the end of branchlets

10/

10/ Tropical conifer
Agathis dammara
Unlike most conifers, which prefer temperate climates, this tree thrives in the humid, tropical forests of Asia. It towers above the canopy on a massive, unbranched trunk and has broad, leathery leaves instead of needlelike foliage. Its cones are nearly spherical and unusually large for a conifer species. When mature, they crumble, releasing papery-winged seeds.

11/ Aerial compost
Platycerium spp.
High in rainforest canopies of Africa, Australia, South America, New Guinea, and Southeast Asia, staghorn ferns grow without soil, clinging to tree trunks as epiphytes. They produce two kinds of leaves: long fronds for photosynthesis and spore dispersal, and flat, overlapping fronds that form a basket. This basket traps leaf litter, creating an aerial compost pile that nourishes the fern and shelters small invertebrates.

12/ Twisted wood
Paullinia spp.
These liana (woody climbers), found in the tropical forests of South America and Africa, show how strange wood can become. Best known for their bright fruits that split open when ripe, the vine stems are also intriguing – often flattened, square, or irregularly lobed. These odd shapes provide flexibility and help plants withstand the shifting, swaying canopy of the forest as they climb towards the sunlight.

13/ Bat bait
Parkia speciosa
In the lowland forests of Southeast Asia, this tall tree produces striking spherical flower heads that dangle from long stalks like golden yellow light bulbs. They are perfectly positioned for bats swooping by at night; as they feed, the bats carry pollen from tree to tree. The flowers are followed by long, strong-smelling bean pods – known as "stink beans".

14/ Bright and bristly
Bombax buonopozense
In the forests of central Africa, the red-flowered silk cotton tree saves its display for the dry season. After dropping its leaves, the bare branches ignite with vivid scarlet blossoms, a beacon to the birds and bats that pollinate them. Below, the plant's massive trunk bristles with thick spines that mount a formidable defence against herbivores.

15/ Starch stash
Metroxylon sagu
A swamp-dweller found in New Guinea's forests, the sago palm grows steadily for about 15 years, storing energy in its trunk. Just before it flowers, the trunk is packed with starch to provide the energy required for reproductive growth. Then, in a single grand effort, it sends up a towering inflorescence, dying soon after. Its scaly fruit has a corky layer within, which makes it buoyant for dispersal by water.

16/ A Jurassic survivor
Dipteris conjugata
Fossils indicate that ferns similar to this plant carpeted Greenland and Europe during the Jurassic Period, when dinosaurs roamed Earth. Today, this fern remains tucked away in the humid forests of tropical East Asia. Its stalkless, broad, fanlike fronds are leathery and deeply lobed to resist wind damage and retain moisture.

TROPICAL AND SUBTROPICAL MOIST FORESTS

11 /

12 /

13 /

14 /

15 /

16 /

PEAS

The third largest family of flowering plants, the pea family (Fabaceae) occurs all over the world except in Antarctica. Its species grow in almost all environments, and range from small herbaceous plants to large trees.

Fabaceae originated along the Tethys sea in the late Cretaceous and diversified in Africa and South America. It dispersed extensively, then different populations adapted to aridification in the Miocene. This resulted in a mosaic of similar-looking trees in at least five genera that all used to be called acacias: *Acacia*, *Acaciella*, *Mariosousa*, *Senegalia*, and *Vachellia*. All family members produce seed pod fruits (legume).

Parkia biglobosa
African locust bean is a large, deciduous tree of the African savanna. It is often grown for its pods, which contain sweet yellow pulp and edible seeds.

Groups of hanging globular flower heads are visited by bats

Compound leaves develop a pair of long spines at their base

Vachellia nilotica
Gum arabic is a large tree widespread throughout the African savanna, the Arabian Peninsula, and the Indian subcontinent. It produces long spines as protection against herbivores.

5 The Mariosousa and Acaciella genera originate from common ancestors with Senegalia and spread throughout Central America and southern North America.

2 Vachellia originates and diversifies in Africa. Long-distance dispersals to South America and India create secondary centres of diversification.

Acaciella angustissima
The prairie acacia differs from most other acacias because it prefers humid places. It behaves like a weedy plant in its native habitat, but it is valued as green manure.

HISTORIC MIGRATION
- Acacia
- Vachellia
- Senegalia
- Other

REPRESENTATIVE LIVING SPECIES
- *Acacia pycnantha*
- *Senegalia greggii*
- *Acaciella angustissima*
- *Vachellia cornigera*
- *Parkia biglobosa*
- *Vachellia nilotica*

FORESTS

Vachellia cornigera
The bullhorn acacia has a symbiotic relationship with ants that live in its hollowed-out thorns. In exchange for a home and sustenance, the stinging ants protect the tree from herbivores.

Small flowers are grouped in elongated heads with numerous yellow stamens

Compound leaves consist of leaflets in opposite rows

❸ Senegalia originates and diversifies in East Africa, in response to climate fluctuations in the Pleistocene. It disperses further in Africa and reaches Asia. It also travels west to South America and on to the southwestern United States.

❹ Acacia disperses further from Southeast Asia and Australia to the Pacific islands up to Hawaii.

Senegalia greggii
The cat claw acacia, recognized by its hooked, clawlike prickles, lives along dry watercourses that fill temporarily. It is a drought deciduous plant that sheds its leaves to conserve water.

❶ Australian Acacia species (wattles) originate and diversify in response to aridification, and then disperse to Southeast Asia and Africa.

Leathery green phyllodes that replace leaves on mature plants perform photosynthesis

Acacia pycnantha
The golden wattle grows in arid regions of southeast Australia and has become naturalized elsewhere. Its phyllodes — flattened stems that function as leaves — are an adaptation to drought conditions.

HEAVY METALS

Humans began mining precious metals in the Neolithic Era. At first, this only involved altering natural habitats on a small scale. However, as mining activities became more sophisticated in the Bronze Age with the extraction of metals from ore via the process of heating and melting (smelting), the impact on plant life became more pronounced. Forests were cleared to supply fuel for fires and to provide structural supports for mines. More significantly, the use of lead and other heavy metals with low melting points in processes to extract metal ore resulted in the contamination of the soil by a variety of heavy metals.

Few plants can survive in soil that is polluted by heavy metals, but a remarkable category of plants known as hyperaccumulators have the ability to either store metals in specialized cell compartments, or excrete them. These species occur naturally on sites where soils contain considerable concentrations of heavy metals and may even indicate the kind of metal that can be found beneath the surface.

As modern mining activities expand, especially in the search for rare-earth elements, the potential for soil pollution becomes a formidable ecological hazard. Hyperaccumulators can play an important role in cleaning polluted soil by extracting heavy metals through their roots and storing them in leaves and stems – a process called phytoremediation.

Hyperaccumulators can also be used to extract metals when they are grown in areas with low concentrations of metal in the soil. In this ecological form of mining – known as phytomining – the plants are harvested and burned to ashes from which metals can be extracted. Although the process is slower than traditional mining, the environmental benefits are significant: no harmful chemicals are involved, and the hyperaccumulators actively restore areas of degraded vegetation.

Blue blooded
Pycnandra acuminata is a small tree found in the mountains of New Caledonia. It is an extraordinary hyperaccumulator that absorbs nickel from the soil and stores it within its sap, causing it to become turquoise in colour.

• FORESTS •

Kapok tree
Ceiba pentandra

Known as "Ceiba" or "Ya'ax ché" in the Yucatec Maya language, the kapok tree (*Ceiba pentandra*) is one of the most culturally significant trees of the Americas. For the Maya, the kapok was the sacred "world tree" – its towering trunk symbolized the connection between the underworld, the earthly realm, and the heavens – and it was often depicted as the cosmic axis linking humans with the divine in Mayan codices (manuscripts) and on carved stone monuments.

Beyond its spiritual importance, the kapok tree has long been valued for its materials. The silky fibres inside its large fruits were traditionally used as stuffing for bedding and pillows, and in the twentieth century, became the buoyant filling for life jackets issued during the Second World War. Its light, workable wood has been carved into canoes, drums, and household objects, while the shade that the tree offers makes it indispensable in cattle pastures across Central and South America, where single specimens are often left standing amid cleared land.

Today, kapok trees remain landmarks in rural communities, used as places for gathering and storytelling. Their immense presence in the landscape continues to inspire reverence, carrying forward a legacy that unites ecological grandeur with deep cultural meaning.

Capsules split open when mature

Seeds are loosely lodged within lightweight fibres

> *"In Cuba, it is revered as magical… even if people are not religious or don't believe, they still never harm the ceiba."*

Euguenia Druyet Zoubareva, "The Sacred Ceiba Tree", 2023

Fruits are long, woody capsules that turn from green to deep brown

Leaves contain mucilage, which is used in traditional medicine

PLANTS OF
TROPICAL AND SUBTROPICAL DRY FORESTS

Tropical and subtropical dry forests occupy the regions around the equator, ranging as far as northern India and southern Bolivia. These regions are warm year-round, but have a strongly seasonal climate, characterized by a long dry season followed by a short, intense wet season. To survive in this challenging environment, plants must be adapted to extremes. Many limit energy expenditure and water loss during the dry season by dropping their leaves and prioritizing flower production. This allows flowers to flourish while the canopy is leafless and improves access for pollinators. Seed is set as the rainy season arrives, improving the chances of germination.

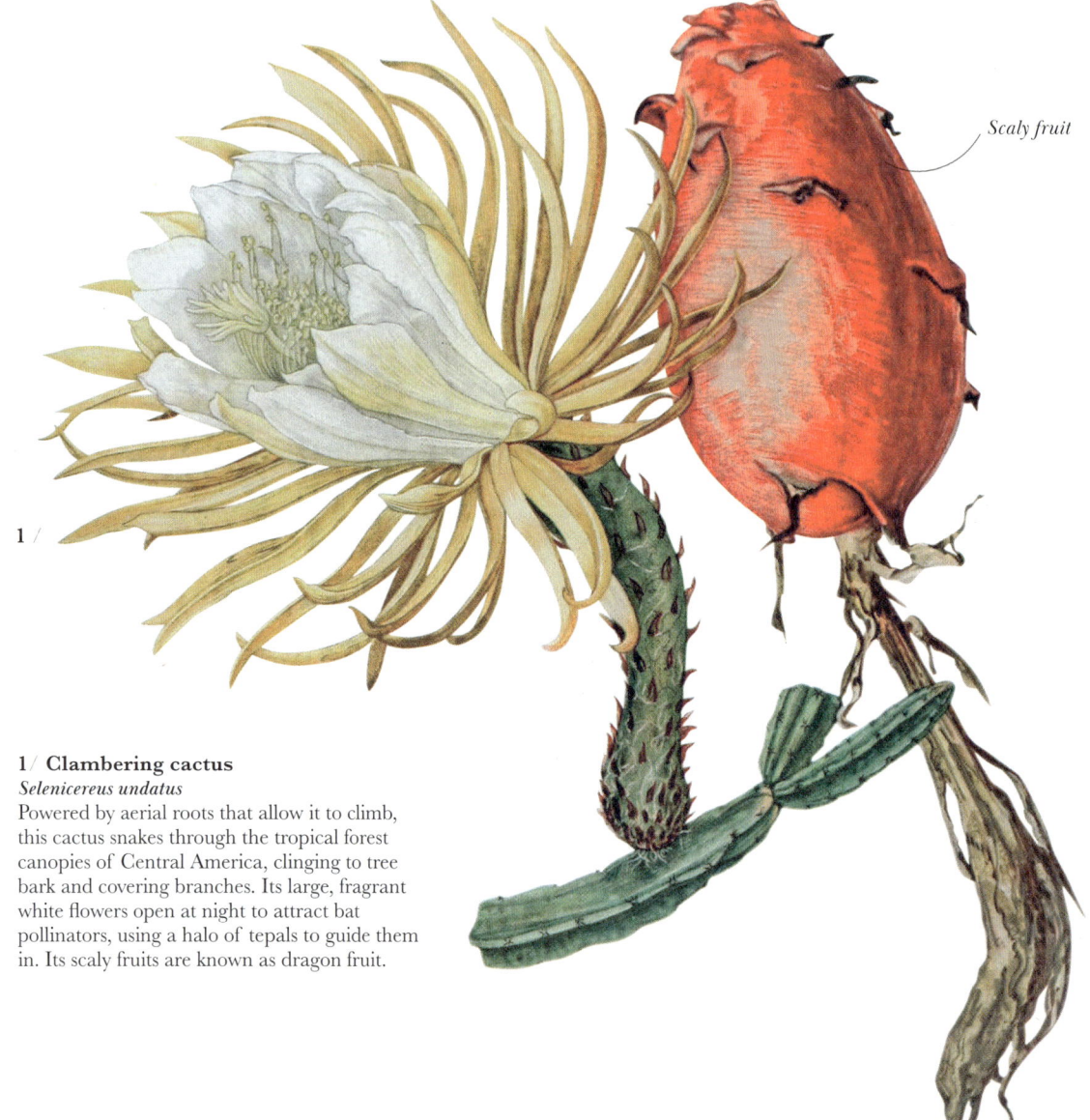

Scaly fruit

1 /

1/ Clambering cactus
Selenicereus undatus
Powered by aerial roots that allow it to climb, this cactus snakes through the tropical forest canopies of Central America, clinging to tree bark and covering branches. Its large, fragrant white flowers open at night to attract bat pollinators, using a halo of tepals to guide them in. Its scaly fruits are known as dragon fruit.

TROPICAL AND SUBTROPICAL DRY FORESTS

2/ Fruitful legacy
Kigelia africana
The sausage tree produces huge fruits that were once consumed by large, now mostly extinct mammals. Today, in its natural habitat of Africa, animals, including elephants and baboons, eat the fruit and excrete the seeds, spreading them over a wide area. In India, where this plant has been introduced and has naturalized, squirrels and termites disperse the seeds.

3/ Floral baubles
Parkia biglobosa
Resembling Christmas tree decorations, the pendent, globe-shaped inflorescences of the African locust bean are made up of small flowers at the end of long stalks. They consist of sterile nectar-producing flowers at the base and fertile flowers above with long, red stamens and styles that give them a festive red colour. These flowers are pollinated by bats or bees.

4/ Curious appetite
Passiflora foetida
The blooms of this stinking passionflower from North, South, and Central America are ringed by three divided bracts that exude sticky digestive enzymes similar to those used by many carnivorous plants. However, it is considered protocarnivorous because while it traps insects to protect itself against damage, it is not known whether it can extract nutrients from prey.

5/ Seeing red
Bixa orellana
The bright red, hairy capsules produced by this South and Central American shrub emerge after a show of large pink flowers. Each capsule contains multiple seeds surrounded by a waxy structure (aril) that is brightly coloured to attract seed dispersers such as birds and ground-dwelling animals.

6/ Landing strip
Delonix regia
This native of dry, deciduous forests in Madagascar is known as the flamboyant tree or flame tree for its showy, bright orange-red flowers. These have five petals, one of which is erect and yellow, or white with red streaks. This special petal provides a landing platform and a nectar guide for pollinating bees, butterflies, and moths.

Landing pad for butterflies

Seed pods follow the flowers

7a /

8 /

9 /

10 /

11 /

• TROPICAL AND SUBTROPICAL DRY FORESTS •

Encouraging cross-pollination
Aristolochia promotes cross-pollination by ensuring that its female structures mature before the male parts. This means that the anthers release pollen only when the flower's female phase is over.

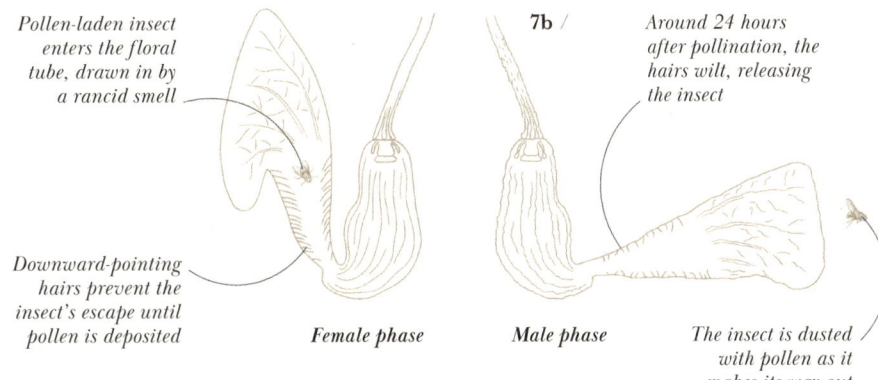

Pollen-laden insect enters the floral tube, drawn in by a rancid smell

Downward-pointing hairs prevent the insect's escape until pollen is deposited

Female phase

7b /

Around 24 hours after pollination, the hairs wilt, releasing the insect

Male phase

The insect is dusted with pollen as it makes its way out

7/ Temporary prison
Aristolochia maxima
This vine from Central and South America has a cunning way to pollinate its flowers. The blooms, which resemble a curved pipe, use a foul odour to lure pollen-carrying insects inside. Here, they are temporarily trapped by downward-pointing hairs, which release them only when they have deposited their pollen gift.

8/ Evolutionary crossroads
Erythrina crista-galli
This bushy shrub or small tree, native to South America, represents an evolutionary transition. Its ancestors used insects for pollination, while fellow *Erythrina* species use birds. This plant makes use of both. Its flowers are brightly coloured and clustered at branch tips to make them easily visible and accessible to hovering hummingbirds. They are also structured to appeal to insects – providing a landing platform and a wider entry point than the narrow blooms of *Erythrina* species solely pollinated by long-beaked hummingbirds.

9/ Arboreal rainmaker
Samanea saman
This tall tree from Central and South America forms a broad canopy, around 30 m (98 ft) wide. It is known as the rain tree because its leaves fold up in response to rainfall and at dusk, allowing raindrops to pass through the canopy, and also because sap-sucking cicadas that feed on its leaves and stems excrete a rain of sticky "honeydew", which falls from the tree.

10/ Floral mutant
Cadia purpurea
This small tree from the dry forests of eastern Africa and Madagascar differs from its relatives in the pea family in having flowers that are radially symmetrical – they radiate from a central point in a regular manner. This mutation helps it to attract a wider variety of pollinators and improves its chances of survival.

11/ A giant of the forest
Ceiba pentandra
This huge tree, up to 70 m (230 ft) high, is widely dispersed in Central and South America, Africa, and parts of Asia. It has large, ribbonlike buttress roots that provide an anchor, especially during high winds. It produces kapok – an extremely light and water-repellent, cottonlike fibre that surrounds its seeds in the pods, and catches the wind to help them disperse.

12/ Forest titan
Tectona grandis
Originating in Southeast Asia, this large tree is surprisingly a member of the mint family, which also includes many low-growing herbs. Its exceptionally hard bark contains high concentrations of oils, resins, and tannins that deter attack from termites and repel water, preventing rot and fungal growth.

Small, white flowers borne in loose clusters above leaves

12 /

13/ Explosive fruit
Hura crepitans
This large tree bears separate male and female flowers. Male flowers are clustered in long spikes. Female flowers are solitary and produce large, pumpkin-shaped capsules that explode when ripe, earning the plant its common name, "dynamite tree". Exploding capsules split into segments and disperse the seeds at speeds of up to 155 kph (96 mph) and as far away as 30 m (98 ft).

14/ Christmas flower
Euphorbia pulcherrima
A member of the spurge family, the poinsettia is cultivated as a popular Christmas pot plant in Europe, but in its natural habitat in Mexico and Guatemala, it displays a shrubby habit. Its festive appearance comes from the large red bracts surrounding tiny inflorescences that consist of naked male and female flowers. The bracts resemble petals to attract pollinators.

15/ Snail-like flowers
Cochliasanthus caracalla
The snail vine belongs to the pea family and is a climbing relative of runner beans. The flowers are asymmetrical, with a coiled stamen tube that resembles a snail shell. As large bees probe for nectar at the base of the standard petal, the style is pushed out against the back of the visitor and dusts it with pollen.

16/ Stuck in a rut
Crescentia cujete
This Central American native bears football-sized fruit with tough, woody shells, housing large seeds. Thought to be adapted for dispersal by Pleistocene megafauna, they have not evolved despite the extinction of these animals. Without natural seed dispersers, the plant has only survived through human intervention, relying on domesticated animals, such as horses and cattle, to eat its fruit and spread its seeds.

17/ The upside-down tree
Adansonia digitata
An enigmatic resident of African dry forests, the baobab is known as the upside-down tree because during the dry season, its bare, sprawling branches looks like roots in the air. During droughts, it loses its leaves and stores water in the tissues of its trunk and branches. As a result, the diameter of its trunk can fluctuate over time.

13 /
Female flower
Emerging male flower
Male flowers grow in conical spikes
Seed capsule
Broadly ovate leaf borne on petiole

• TROPICAL AND SUBTROPICAL DRY FORESTS •

14 /

16 /

17 /

15 /

Lophostemon confertus
Brush box is a large, highly adaptable tree from Eastern Australia. It is a dominant species in moist forests with variable rainfall, and it is often planted as a street tree.

Fruit is a woody, bell-shaped capsule

Aromatic foliage contains volatile oils that repel some insects

8 The genus Myrtus possibly disperses over long distances across Africa and establishes in the Mediterranean.

5 Long-distance dispersal from Australia and New Zealand to the Pacific Islands and western South America creates distinct relict populations.

7 Mainly fleshy fruited species disperse and diversify further through the neotropics and the southern part of North America.

6 Fleshy fruited species are dispersed over long distances from Africa to South America likely with the help of birds and bats.

NORTH AMERICA

PACIFIC OCEAN

SOUTH AMERICA

AFRICA

MYRTLES

The largely evergreen myrtle family (Myrtaceae) has a widespread distribution across the Southern Hemisphere, mainly in wet or dry tropical and subtropical forests.

Myrtaceae originated in the Southern Hemisphere's supercontinent of Gondwana more than 90 million years ago. As the continent broke apart, its populations became isolated, driving diversification in Africa, Southeast Asia, and South America, as well as secondary colonizations in Europe and South America. All species in the family produce essential oils – notably eucalyptus, tea tree, and cloves.

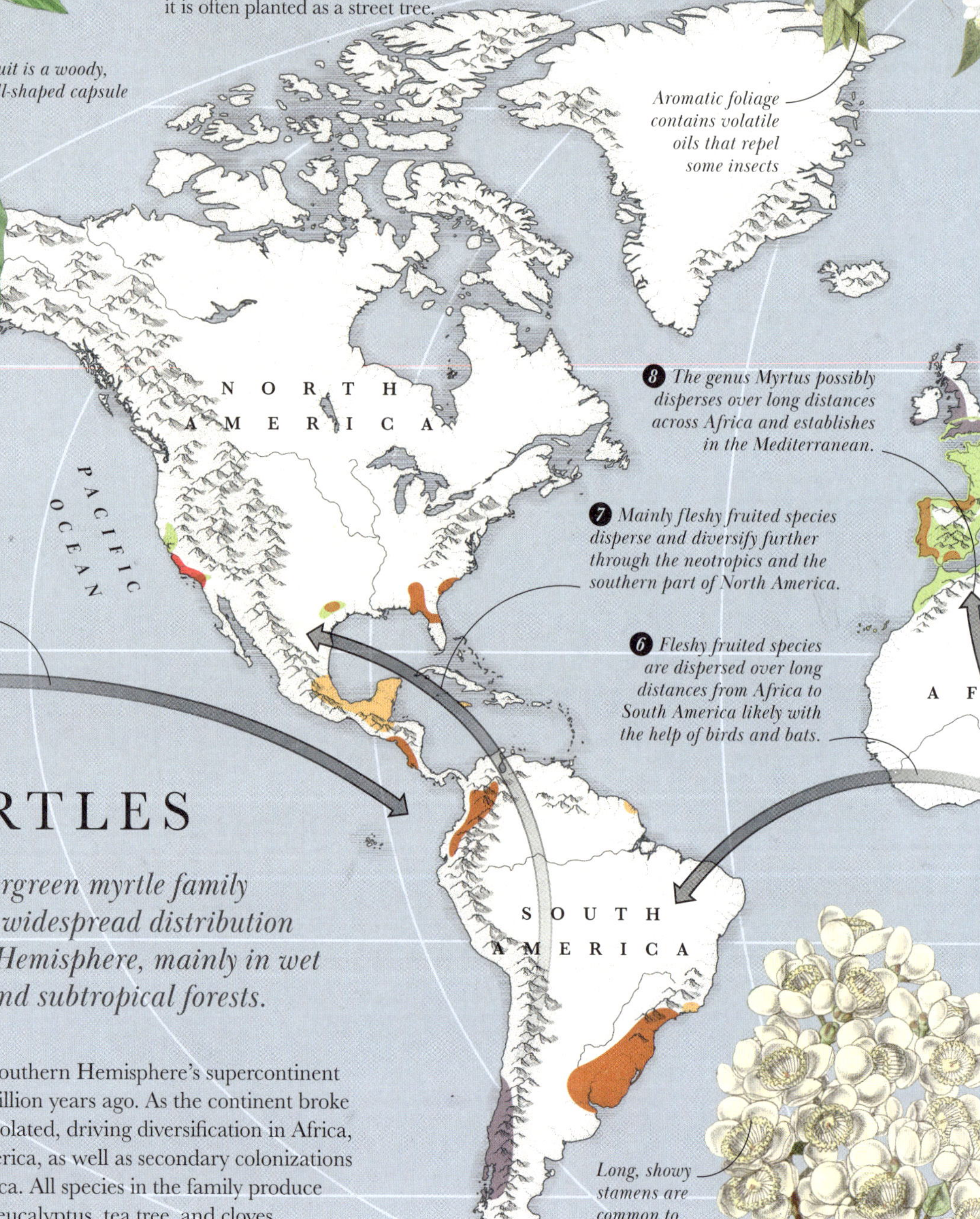

HISTORIC MIGRATION
Myrtaceae ➤

REPRESENTATIVE LIVING SPECIES
- *Feijoa sellowiana*
- *Lophostemon confertus*
- *Luma apiculata*
- *Myrtus communis*
- *Pimenta dioica*
- *Syzygium aromaticum*

Long, showy stamens are common to this group

Luma apiculata
This small tree is native to the central Andes. It forms dense evergreen clumps along streams, and leaves almost no space for other tree species.

• FORESTS •

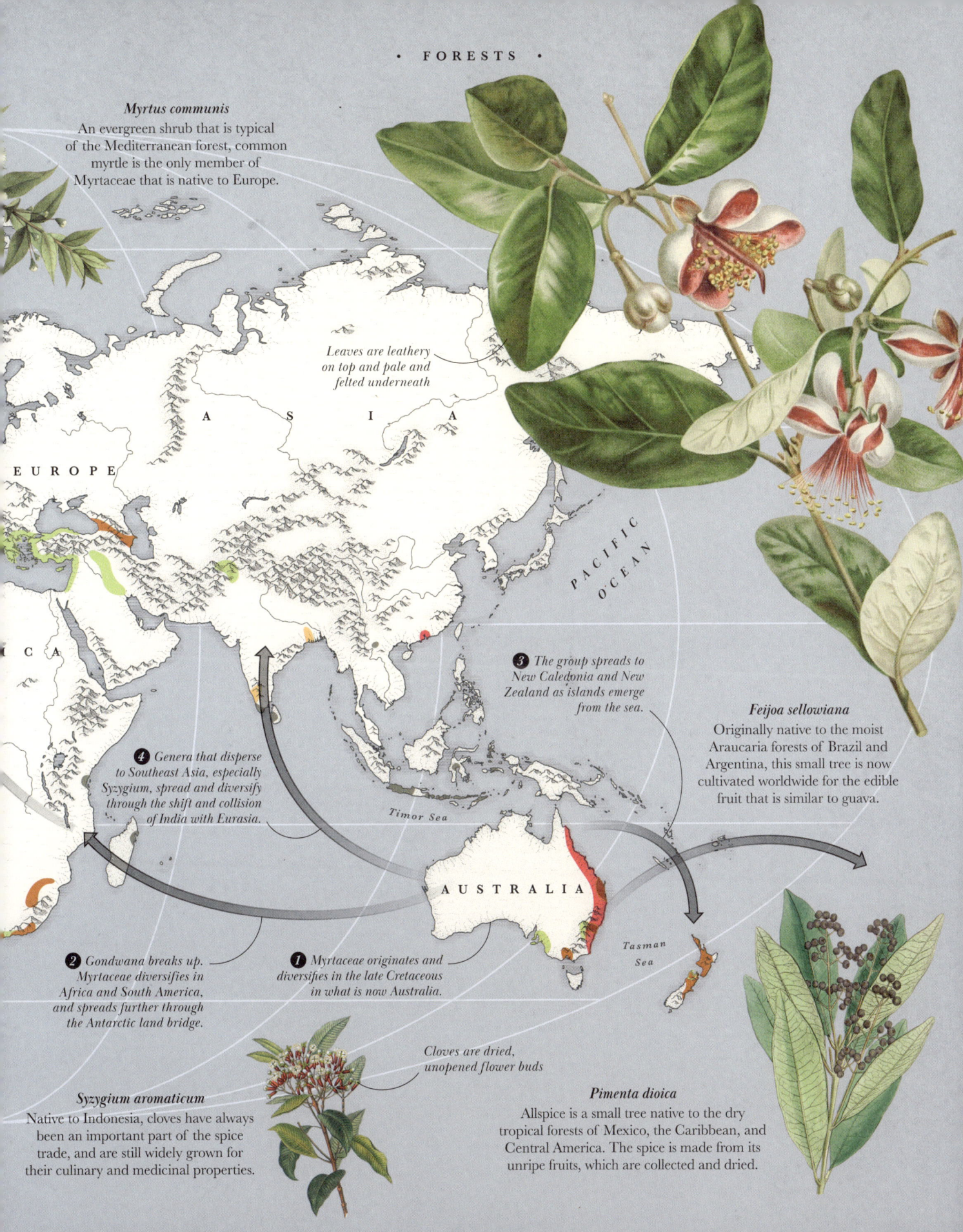

Myrtus communis
An evergreen shrub that is typical of the Mediterranean forest, common myrtle is the only member of Myrtaceae that is native to Europe.

Leaves are leathery on top and pale and felted underneath

❸ The group spreads to New Caledonia and New Zealand as islands emerge from the sea.

Feijoa sellowiana
Originally native to the moist Araucaria forests of Brazil and Argentina, this small tree is now cultivated worldwide for the edible fruit that is similar to guava.

❹ Genera that disperse to Southeast Asia, especially Syzygium, spread and diversify through the shift and collision of India with Eurasia.

❷ Gondwana breaks up. Myrtaceae diversifies in Africa and South America, and spreads further through the Antarctic land bridge.

❶ Myrtaceae originates and diversifies in the late Cretaceous in what is now Australia.

Cloves are dried, unopened flower buds

Syzygium aromaticum
Native to Indonesia, cloves have always been an important part of the spice trade, and are still widely grown for their culinary and medicinal properties.

Pimenta dioica
Allspice is a small tree native to the dry tropical forests of Mexico, the Caribbean, and Central America. The spice is made from its unripe fruits, which are collected and dried.

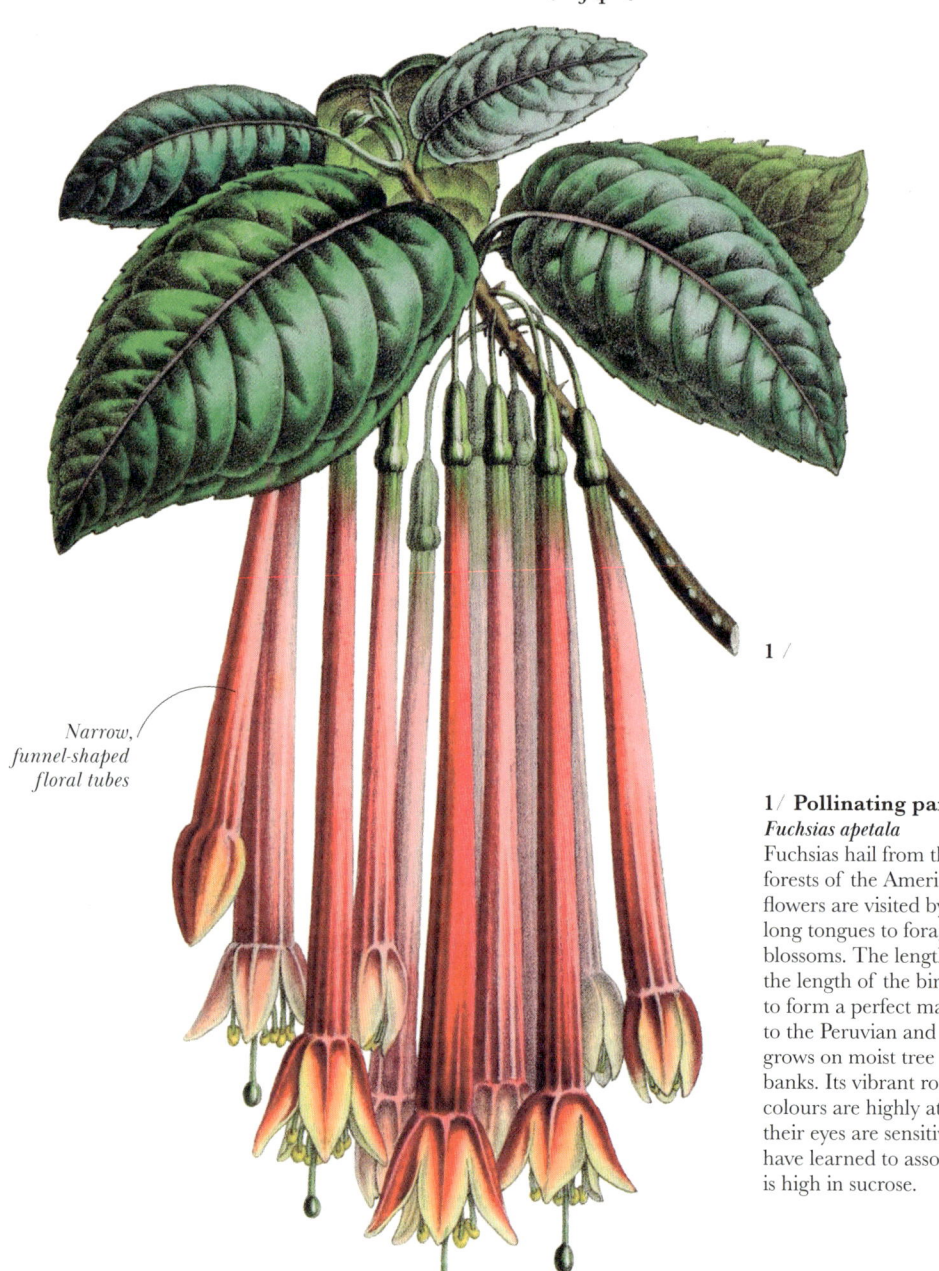

Narrow, funnel-shaped floral tubes

1/ Pollinating partner
Fuchsias apetala
Fuchsias hail from the high-elevation cloud forests of the Americas, where their pendant flowers are visited by hummingbirds who use long tongues to forage nectar from the tubular blossoms. The length of the flower tubes and the length of the birds' beaks have coevolved to form a perfect match. *Fuchsia apetala* is native to the Peruvian and Bolivian Andes where it grows on moist tree trunks, rocks, or sphagnum banks. Its vibrant rose-red and orange-red colours are highly attractive to hummingbirds – their eyes are sensitive to these hues, which they have learned to associate with nectar that is high in sucrose.

PLANTS OF THE

CLOUD FOREST

Cloud forests occur on tropical mountains across Africa, Asia, and the Americas, where persistent mist blankets the slopes in cool, wet air. These high-elevation habitats are defined by constant moisture, frequent rainfall, and often steep terrain. Plants here must adapt to limited sunlight, saturated soils, and fluctuating temperatures. The result is a community rich in epiphytes, parasites, and specialized trees – figs that strangle their hosts, towering palms, and orchids that photosynthesize with their roots – all shaped by the cloud-soaked environment they inhabit.

CLOUD FOREST

2/

3/

4/

2/ Massive moss
Dawsonia superba

Mosses generally lack the complex transport tissues found in the stems of vascular plants, which explains in part why they are so much smaller. The *Dawsonia* genus, found in the cool humid cloud forests of Southeast Asia, Australia, and New Zealand, is one exception. Its species have evolved specialized tissues similar to those of vascular plants to transport water and nutrients. This is what has enabled *Dawsonia superba* to become the world's tallest self-supporting moss, reaching heights of 60 cm (24 in).

3/ Evolutionary reversal
Drimys granadensis

This evergreen shrub grows up to 12 m (40 ft) tall, primarily in cloud forests from southern Mexico through to Peru. Its large, star-shaped flowers are pure white, fragrant, and larger than those of other *Drimys* species. Unlike most flowering plants, members of the *Drimys* genus lack xylem vessels – specialized cells that transport water from the roots to the rest of the plant. Instead, the plant uses more primitive elongated conducting cells called tracheids to transport water. This adaptation is thought to help the plant withstand the freezing temperatures of high-altitude cloud forests. It is a trait that characterizes the Winteraceae family, which diverged from other plant lineages in the early Cretaceous Period.

4/ Silvery strands
Tillandsia usneoides

Spanish moss – which confusingly is neither a true moss, nor from Spain – is an air plant that drapes trees in silvery garlands of shoots. It is widely distributed across the subtropics and tropics, including cloud forests in the Americas. Like other air plants, Spanish moss absorbs water from the mist, using tiny leaf scales to capture droplets and channel them inside.

5/ Healing bark
Cinchona pubescens

Native to Central and South America, *Cinchona pubescens* is also called the red cinchona because its bark can turn reddish when cut. The red cinchona can recover from serious damage, regrowing its bark and sending up new stalks after the tree has been felled. As well as helping the tree heal, this special bark has proved invaluable in healing humans. It contains a special alkaloid, quinine – an active compound that proved to be one of the first effective medications to combat malaria. It also has anti-inflammatory properties and is effective against fevers and muscle cramps.

5/

Pink, fragrant flowers

6 /

8a /

9 /

7 /

10 /

6 / Colombian colossus
Ceroxylon quindiuense
Native to Colombia and Peru's cloud forests, the Quindio wax palm is the world's tallest palm, with trunks soaring more than 60 m (200 ft). In the past, it suffered a severe decline because the protective wax that coats its leaves was scraped off and used to make candles. Today, it is appreciated not only for its significant stature, but also as a vital nesting habitat for the endangered yellow-eared parrot, which relies on the palm for food and shelter.

7 / Lofty waterhole
Guzmania conifera
Many members of the bromeliad family, including *Guzmania*, form rosettes that collect rainwater, creating miniature ponds high in the forest canopy that function as their own tiny ecosystems. *Guzmania conifera* has a distinctive central flowering spike formed of overlapping bracts that shine bright red, their edges tipped with golden-yellow. It is an epiphyte that grows on other plants, but it only uses them for support, not for sustenance. Instead, tiny hairs on its leaves absorb both moisture and nutrients directly from the air.

8 / Security system
Cecropia peltata
Hallmarks of tropical forests in the Americas, *Cecropia* trees have distinctive umbrella-like leaves and hollow stems. Fast-growing, they quickly colonize landslides and forest gaps. Their survival is tied to three aggressive ant species of the genus *Azteca*, which nest inside the stems and patrols the tree, attacking and driving away herbivores and insects. If the tree is wounded, the ants quickly repair the damage, using sap and plant fibres to patch up any holes. The ants also kill the ends of vines attempting to grow up the trunk to prevent them smothering the tree. In return for these defence services, *Cecropia* offers its protectors shelter and a source of glycogen-rich food, produced at the base of its leaves.

9 / Forest architect
Ficus tuerckheimii
Strangler figs begin life as epiphytes high in the tree canopy, sending roots down to the ground while weaving a lattice of branches around their host. The host gradually dies, leaving the fig as a hollow cylinder of fused wood. These towering structures provide both food and shelter for countless forest animals.

10 / Sticky seeds
Psittacanthus cucullaris
High in the canopy of Central and South American cloud forests, parrot flowers thrive as hemiparasites, drawing water and nutrients from host trees while photosynthesizing with their own leaves. Their vivid flowers attract pollinating hummingbirds and fruits entice other birds to visit and spread their seeds. Adapted for this lifestyle, the seeds are coated in sticky tissue so they adhere to branches after passing through the digestive systems of birds.

11 / Habitat hallmark
Weinmannia pubescens
Found in Venezuela, Colombia, Peru, and Ecuador, this plant belongs to a genus of 90 species of trees and shrubs native to cloud forests in North, Central, and South America, and the Caribbean. Weinmannias are so characteristic of cloud forests that they act as an indicator species for this unique habitat. They have distinctive leaves, with leaflets arranged in opposite pairs, and prominent stipules (leaflike appendages) on the stalks that join the leaves to the stems. *Weinmannia pubescens* plays an important role as a pioneer species: it has been found regenerating disturbed areas, such as hillsides affected by landslides.

Small, white or cream flowers grow in dense clusters

11 /

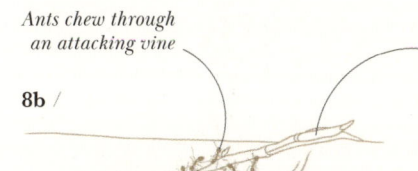

Ants chew through an attacking vine

8b /

Stem at risk of strangulation

Ants on patrol
Ants mount a 24-hour patrol on Cecropia stems. When a threat is identified, reinforcements are called to tackle the danger.

12/ Flamboyant flower
Anthurium scherzerianum

There are over 1,000 species of anthuriums, most of which are epiphytes clinging to trees in cloud forests. Their vivid, colourful display comes from a spathe – a bright bract – that frames the spadix, a spike packed with tiny flowers. One species, *Anthurium scherzerianum*, native only to Costa Rica, has an unusual curly spadix. This almost necklike feature, along with the bright red or pink colour of its spathe, has earned it the evocative common name flamingo flower.

13/ Landing lights
Columnea consanguinea

An epiphyte of cloud forests, this shrub-like herb is famous for the translucent red spots on its leaves that glow when backlit by the sun. Unlike its more showy relatives, its flowers are modest, so it uses these glowing red cues to attract hummingbirds, which home in on the patterns as reliable guides to nectar rewards.

14/ Unlikely companions
Cora glabrata

Lichens are complex organisms that are essentially the product of a mutually beneficial relationship between a fungus and an alga. While most lichens unite algae with a specific group of fungi, known as sac fungi, the *Cora* genus is unusual. It belongs to a small set of basidiolichens, where the fungal partner is a basidiomycete – the group that includes mushrooms and bracket fungi. This explains *Cora glabrata*'s mushroomlike form on the cloud forest floor of its American habitat.

15/ Subterranean parasite
Langsdorffia hypogaea

Native to Central and South America, Madagascar and Papua New Guinea, this subterranean parasite of oak trees is one of the most unusual plants of the cloud forest. Leafless and without chlorophyll, it remains underground as a complex system of tubers connected to the roots of its host until it flowers and fruits. The remarkable inflorescences are often mistaken for fungi and sometimes likened to deep sea creatures.

16/ Versatile roots
Taeniophyllum sulawesiense

Among the most unusual orchids of cloud forests are approximately 250 species of *Taeniophyllum*, found from Africa to the western Pacific. Commonly known as ribbon roots, these tiny epiphytes have dispensed with leaves altogether. Instead, their roots are flattened, green, and photosynthetic, clinging to tree bark like ribbons. This radical shift allows the plant to capture light and water efficiently in the cloud forest canopy.

17/ Canopy carnivore
Nepenthes rafflesiana

In the cloud forests of Asia, carnivorous *Nepenthes* plants solve nutrient shortages by digesting insects, luring them in via their remarkable leaf traps. Each leaf tip extends into a tendril that bears a fluid-filled pitcher, often shaded by a lid; insects tumble in and are consumed. *Nepenthes rafflesiana*, widespread across Southeast Asia, is special in having two types of trap: its narrow upper pitchers have a distinctive raised lip and target flying insects, while wider pitchers lower down its stems attract crawling insects. Some plants can produce pitchers up to 35 cm (14 in) long.

Distinctive curly spadix is likened to a pigtail

12/

CLOUD FOREST

13 /

14 /

15 /

16 /

17 /

FIGHTING DISEASE

A tall, elegant feature of woodlands, hedges, and gardens, the European ash (*Fraxinus excelsior*) is widely distributed across almost all European countries. Since the 1990s a deadly disease, called ash dieback, has devastated whole populations, which appeared to have little or no natural defence against the disease.

The disease is caused by the fungus *Hymenoscyphus fraxineus*, which blocks the water transport system in the tree, causing leaves to wilt and diamond-shaped lesions to form where its branches meet the trunk. The fungus is thought to have been introduced to Europe from Asia, either spread by the wind or accidentally imported on trees that were already infected. It has since decimated ash woodlands in large areas of mainland Europe and estimates suggest that ash dieback could kill up to 80 per cent of ash trees in the UK.

The loss of mature trees could have a devastating impact on the landscape and the biodiversity of woodlands in a similar way to Dutch elm disease – another fungal disease that killed millions of elms in the 1970s. However, unlike Dutch elm disease, ash dieback progresses slowly and a small number of ash trees now appear to show some resistance to the infection – so they may eventually recover over time. Ash trees also produce an abundance of seedlings and there are grounds for hope that some of these have developed sufficient genetic variation to make them fully resistant to the disease. The careful management of ash populations – by identifying and propagating trees that are more immune to infection – and the ash tree's ability to adapt via natural selection, may help this tree to overcome a pernicious disease.

Top-down attack
The bare branches of infected ash trees near Littondale in the North Yorkshire Dales, UK, show how the trees die from the top downwards. Only a few of the lower leaves on some of the trees still have access to water.

FORESTS

PLANTS OF THE
MEDITERRANEAN FOREST

Mediterranean forests and scrublands occur at mid-latitudes, between 30 and 40 degrees north and south of the equator and include five regions: the Mediterranean basin, chaparral in California, the Chilean matorral, fynbos in South Africa, and mallee woodlands of South West Australia. The forests are found in areas characterized by an alternation of wet, mild winters and hot, dry summers. Fires are a regular occurrence and so many plants in these forests have developed a resistance to fire or have regenerative strategies to recover from fire damage. Most Mediterranean forests have been degraded as a result of excessive logging, fire, and overgrazing.

1 /

2 /

3 /

4 /

1 / Regulating growth
Araucaria araucana
The monkey puzzle forms extensive forests in Chile and Argentina. The species produces separate male and female plants and studies of tree-ring patterns has revealed differences in growth patterns between the two sexes. A year after pollination, the female tree's growth slows considerably as it puts its energy into developing its seed cones. During a pollination year, it is the male's growth rate that slows as it reallocates its resources to produce pollen cones.

2 / Regeneration specialist
Arbutus menziesii
Its ability to rapidly resprout after a fire via bulbous structures called lignotubers (see *Sequoia sempervirens*, p.105) allows the Pacific madrone to act as a nurse tree. In this role it provides shelter for smaller, slower-growing trees and so aids forest regeneration. The madrone also hosts a high diversity of mycorrhizal fungi, which enhance nutrient absorption. This improves the forest's resilience to disturbance. By contrast, it is thought that its thin, peeling bark contains chemicals that inhibit germination and growth of other plants, which has a very different effect, eliminating competitors that lack the madrone's regenerative powers.

3 / Temporary prison
Arisarum vulgare
Friar's cowl grows in shady spots in the forests of the Mediterranean basin. Its inflorescence consists of a striped tubular spathe surrounding a clublike spike (spadix), which bears a succession of male and female flowers near its base. Attracted by its scent, flies crawl into this tube and become temporarily trapped, pollinating the flowers as they try to escape.

4 / A toxic alliance
Nerium oleander
Growing along streams in the Mediterranean basin, this tree is highly toxic, bitter, and unpalatable for most. Its pollinators include the oleander hawk-moth (*Daphnis nerii*) and spotted oleander caterpillar moth (*Empyreuma pugione*). These moths lay their eggs on the leaves of the oleander so the larvae have a ready supply of food once they hatch. The caterpillars are immune to the plant's toxins and so can feed on the leaves, which in turn makes them toxic to potential predators.

• MEDITERRANEAN FOREST •

5/ Invasive beauty
Acanthus mollis
Native to the southwest Mediterranean, bear's breeches grows in shady spots in open woodlands. Its pollinators include large insects, mostly bumblebees strong enough to push and navigate their way through its hooded flower, past its two pairs of robust stamens, to the nectar. Its rhizomes spread aggressively, regenerating from the smallest fragment, which allows it to thrive in disturbed ground.

6/ Emblematic tree
Cedrus libani
The national emblem of Lebanon, the cedar of Lebanon grows at altitudes of roughly 1,300–3,000 m (4,270–9,850 ft) in the Eastern Mediterranean basin. While this resilient tree is drought-tolerant and frost-resistant, it flourishes in cool, moist conditions. Its durable heartwood resists decay, needle-like leaves reduce water loss, and deep roots help to stabilize soil.

Female cone is larger and egg-shaped

Male cone is small and cylindrical

Hoodlike purple sepals and bracts protect the flowers

Dark green, deeply lobed leaves inspired the design of Corinthian columns

7 /

8 /

9 /

10 /

11a /

12 /

• MEDITERRANEAN FOREST •

7/ Far from home
Erica arborea
A medium-sized shrub of the Mediterranean basin, tree heather grows in dense maquis scrubland or forests. It has a discontinuous distribution across the Mediterranean and is also found in East Africa as the result of an expansion of its range during the Pleistocene. Its white, bell-shaped blooms flower from late winter to spring, ahead of the dry summers, while there is sufficient water supply. Its needlelike leaves reduce water loss.

8/ Floral mimicry
Ophrys speculum
Exclusively pollinated by the *Dasyscolia ciliata* wasp, the mirror orchid's flowers have evolved to look and smell like a female wasp. Each flower comprises a shiny lip or labellum, with hairs along its edge, to resemble the female wasp. This fools male wasps into trying to copulate with the flowers. In doing so, they pick up pollen from one flower and deposit it on the stigma of another, fertilizing it.

9/ Seeded competitor
Smilax aspera
A low-growing climber, common smilax is a typical plant of the forest understorey in the Mediterranean basin. The number of fertilized seeds in its fruit – a round berry – varies from one to three, with differences in seed size being inversely proportional to their number. In areas where competition from similar fruit-bearing plants is greater, the plant produces berries with single seeds as they have a higher pulp-to-seed ratio to attract birds for seed dispersal.

10/ Sturdy seeds
Ceratonia siliqua
Native to the Mediterranean region, Carob is also extensively cultivated there for its edible seeds, which are used as animal feed, and its pulp, which is used as a substitute for chocolate. A member of the pea family, its seedpods grow directly from the branches or from the tree trunk, a phenomenon known as cauliflory. The small flowers are unisexual or sometimes hermaphroditic and are pollinated by both insects and wind. The ripe, pulpy pods are eaten by animals and the hard seeds, which travel through their digestive tracts intact, are dispersed in the excrement.

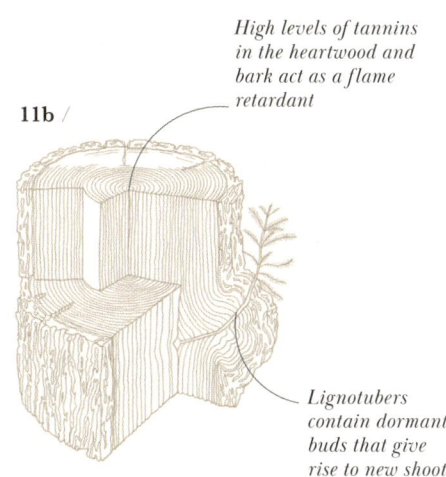

11b /

High levels of tannins in the heartwood and bark act as a flame retardant

Lignotubers contain dormant buds that give rise to new shoots

Regenerative redwood
Coast redwood has a remarkable regenerative capacity. It can sprout a new plant from structures called lignotubers that grow at the base of the tree, usually below the soil line.

11/ A champion tree
Sequoia sempervirens
Confined to a narrow coastal fog belt in the US, coast redwoods can grow up to 60 m (197 ft) in a century under favourable conditions. They are among the world's fastest growing and tallest trees reaching heights in excess of 105 m (345 ft). Their long life is thanks to their resistance to fire and disease, with some trees being more than 2,000 years old.

12/ A giant among palms
Jubaea chilensis
Endemic to the Chilean matorral, the Chilean wine palm is a cold-tolerant, non-tropical palm and the only palm native to Chile. This slow-growing, enigmatic giant can grow more than 24 m (80 ft) tall with a trunk up to 1.5–2 m (5–6½ ft) wide. Its native population has suffered a significant decline due to overharvesting for its sweet sap, resulting in few remaining stands and restricted genetic diversity.

13/ Specialized roots
Embothrium coccineum
This large shrub with bright red flowers grows in the Chilean matorral. Seedlings form cluster (proteoid) roots. These short rootlets form a thick mat beneath the leaf litter that enhance nutrient uptake and allow access to otherwise inaccessible forms of various nutrients, especially phosphorus. This enables plants to grow in poor, nutrient-deficient soils.

13 /

Brightly coloured flowers attract bird pollinators

14/ Off-putting taste
Quillaja saponaria
Soapbark is the only member of the family Quillajaceae native to central Chile. It is named for the high levels of saponins found in its bark, a soapy compound that foams when mixed with water. Saponins are bitter tasting and toxic, which works to discourage herbivory. Soapbark's star-shaped flowers develop into pinwheel-shaped fruits with five follicles, or sections, each housing about 20 seeds. The fruit matures to dry brown and splits open to release tiny, winged seeds that disperse on the wind.

15/ Pine lookalike
Allocasuarina fraseriana
The western sheoak is endemic to South West Australia, growing in dry forests or on coastal shorelines. Despite being a flowering plant, it bears a strong resemblance to conifers having adapted to drought in similar ways, though completely independently. The leaves are reduced to tiny scales on evergreen shoots that look like pine needles. The female flowers develop into cones resembling those of conifers.

16/ Double parasitism
Cytinus hypocistis
This small parasitic plant from the Mediterranean is mainly found growing on the roots of *Cistus*, or rockrose, plants. It lacks any chlorophyll for photosynthesis and forms a three-way association with its host and mycorrhizal fungi – fungi that penetrate the roots of other plants. In doing so, the parasite avoids overexploiting the rockrose's resources. It appears above ground during blooming season as a cluster of fleshy stems covered in red scales, bearing yellow flowers.

17/ Umbrella-shaped canopy
Pinus pinea
A large tree of the Mediterranean basin, the parasol pine has a flattened, umbrellalike canopy. Unlike most other pines, its seeds have reduced wings that do not allow for wind dispersal. Originally distributed by Iberian magpies, these seeds are now mostly dispersed by humans who have taken to cultivating this tree on a large scale for its edible seeds, known as pignon or pine nuts.

18/ Ant-aided dispersal
Cyclamen hederifolium
This small herbaceous plant is native to the Mediterranean basin. Its seeds are specialized for dispersal by ants, something called myrmecochory. The ants are after the fleshy edible aril on the seeds. Once the flowers are pollinated, the flower stems coil, dragging the seed capsules to the ground and enabling the ants to gather the seeds when the capsule opens.

19/ Chemical defence
Umbellularia californica
The Californian laurel is an evergreen tree of the mixed evergreen forests of California, US. Its leaves contain the compound umbellulone, which affords it a range of chemical defences. When released into the soil, it can inhibit the growth of competing plants. Its bitter taste deters insects and herbivores from damaging the leaves, while its pungent smell repulses most browsing animals and can cause headaches in humans. Small, fragrant flowers appear in late winter or early spring, before new leaves can come in and mask their scent.

20/ Leaflike stems
Ruscus aculeatus
Butcher's broom is found in dry shaded woodlands of the Mediterranean region, extending to western Asia. What appear to be leathery leaves adapted to drought are flattened stems called cladodes or phylloclades, as the original leaves have become reduced to scales and are non-photosynthetic. The "leaves" produce small white flowers on their underside, which develop into red berries.

14 /

Fruit capsule persists on the plant after splitting open

• MEDITERRANEAN FOREST •

15 /

19 /

16 /

18 /

20 /

17 /

Clerodendrum trichotomum
A large deciduous shrub native to subtropical and temperate east Asia, harlequin glorybower grows on low mountains below 2,400 m (7,870 ft). It is cultivated, but considered invasive in some areas.

White flowers grow in clusters at the end of the stems

Large, fragrant flowers mature into drupes with a red calyx and blue fruit

4 Some genera arrive in North America from East Asia via the Bering Sea land bridge about 10 million years ago, and diversify further.

8 Secondary centres of diversity for Salvia and other genera are established, especially in Mexico, as the family continues to expand throughout the Americas.

7 Repeated long-distance dispersals of Salvia and other genera from the Mediterranean to South America lead to the establishment of many new species.

MINTS

The mint family (Lamiaceae) occurs worldwide, except in the far north. Many of its members bear opposite leaves that are often aromatic, and some, such as sage and mint, are familiar culinary herbs.

Lamiaceae originated in Southeast Asia around 65 million years ago as woody plants that spread rapidly, with repeated transitions between woody and herbaceous habits. Some woody species such as *Prostanthera* migrated to Australia and others remained in Southeast Asia, while climatic changes caused by mountain building drove the distribution of herbaceous species into Europe, Africa, and the Americas.

Flowers have a lever mechanism that presses pollen onto the backs of visiting bees

HISTORIC MIGRATION
Lamiaceae →

REPRESENTATIVE LIVING SPECIES
- Clerodendrum trichotomum
- Mentha longifolia
- Prostanthera cuneata
- Salvia officinalis
- Teucrium chamaedrys
- Vitex agnus-castus

Salvia officinalis
A subshrub of the dry, sunny maquis in the Mediterranean region, common sage has also become naturalized in various temperate parts of the world.

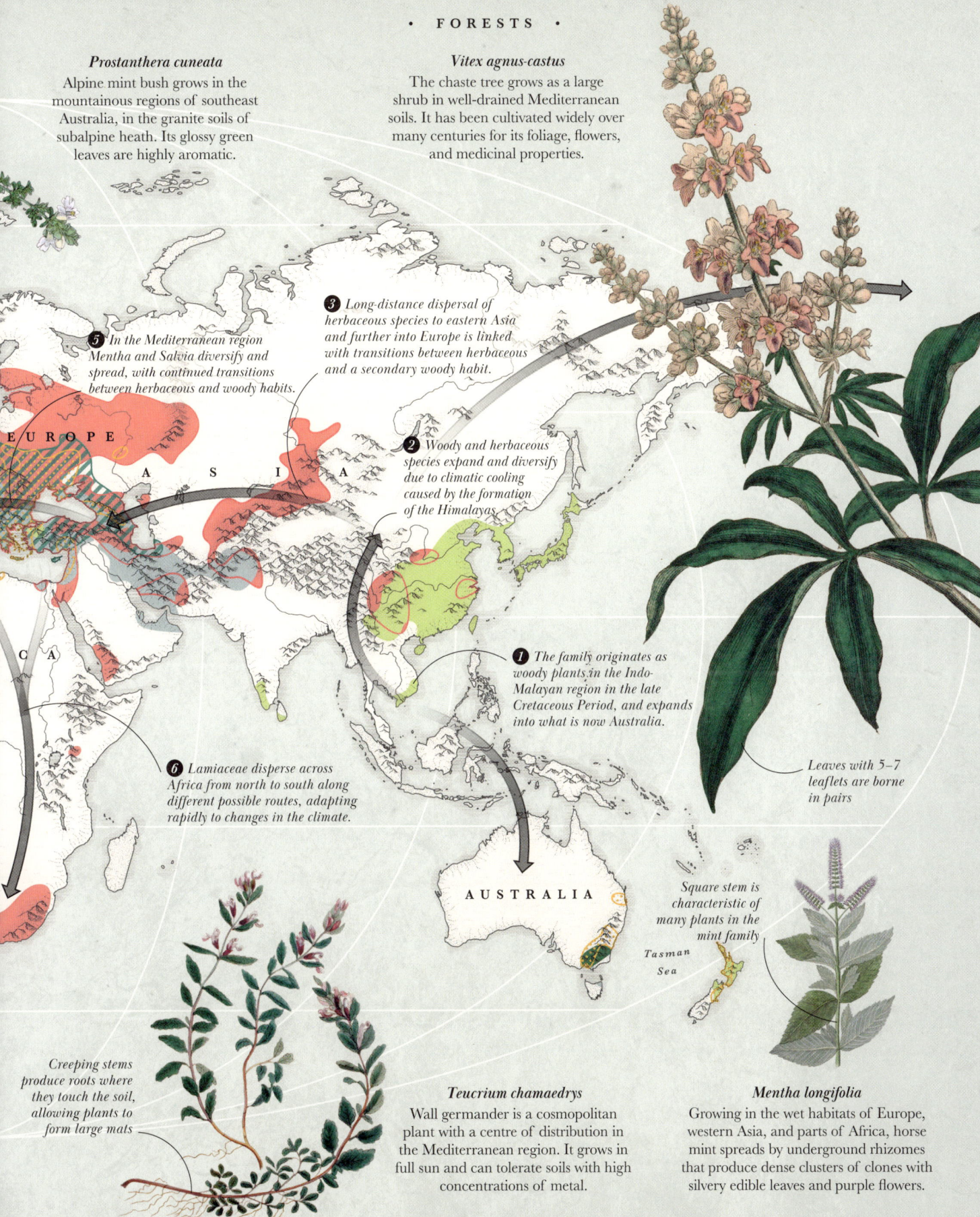

PLANTS OF THE
TEMPERATE FOREST

Sweeping across parts of eastern North America, West and Central Europe, East Asia, and Chile, temperate forests enjoy a seasonal climate, with short winter days and low temperatures giving way to longer, hotter days in summer, and with moderate rainfall all year round. As this biome encompasses around 25 per cent of the world's forests and is spread across a range of continents and latitudes, it is broken down into three main categories: deciduous forest, dominated by large, broadleaved trees; coniferous forest, largely containing needle-leaved evergreens; and mixed forest, featuring a combination of deciduous trees and conifers.

Red-purple fringe makes the flower more visible in low light, and guides bees to nectar rewards

1 /

1 / Welcome haven
Helleborus foetidus
Widespread in Europe, this hellebore is one of the first plants to flower each year, providing vital food for early-rising queen bumblebees. Many other bee species also receive a warm welcome – the plant's nectaries (nectar-producing organs) are colonized by yeasts that give off heat as they metabolize nectar sugars. This makes the flowers up to 2°C (36°F) warmer than the surrounding air, attracting heat-loving bees in the cool of spring.

2 / Flamboyant foliage
Acer palmatum
An important constituent of temperate forests in Japan and Korea, the Japanese maple is a relatively large understorey tree with delicate hand-shaped leaves. The genetics of the tree determine the colour of its leaves by dictating the ratio of different leaf pigments. Some trees naturally produce high levels of anthocyanins, responsible for red colours, giving them red leaves in spring; in other trees, the green pigment chlorophyll dominates, so they produce green leaves. The pigment carotenoid, which creates yellow and orange colours, is present in all leaves, but only reveals itself in autumn when the temperatures begin to cool and chlorophyll starts to break down. This is what helps to create a spectacular display of autumnal colours.

2 /

Palm-shaped serrated leaves contain five to seven lobes

• TEMPERATE FOREST •

3/ **Long-lived lily**
Convallaria majalis
Native to shaded woodland in Europe and East Asia, lily of the valley produces small, sweetly scented white flowers that appear in late spring. It propagates mainly through horizontal runners that extend from the base of the plant and gradually form extensive colonies, some of which are thought to be up to 600 years old. This slow spread over generations means that lily of the valley can be used as a useful bioindicator to identify areas of ancient woodland. Despite its delicate beauty, all parts of the plant are highly poisonous.

4/ **Rapid riser**
Liriodendron tulipifera
Known as the tulip tree for its distinctive cup-shaped, butter yellow flowers, this native of eastern North America often forms groves in moist, well-drained soils. A "pioneer" species, it thrives in newly created or damaged environments, such as habitats disturbed by fire, storms, or logging. Its extensive root structure helps to stabilize the soil, allowing other plants to establish themselves. It can reach heights over 46 m (150 ft), and its rapid growth quickly provides a canopy – turning butter-yellow in autumn – for a variety of shade-loving understorey plants.

5/ **Master of self-defence**
Toxicodendron radicans
Poison ivy is extremely adaptable, growing as a woody shrub, a crawling vine, and a climbing vine in many habitats in North America. Its leaves, stems, and roots contain a highly toxic oily resin called urushiol, that protects the plant from predators and pests, and also forms a hard lacquer that prevents pathogens from entering through the bark.

6/ **Determined settler**
Reynoutria japonica
Japanese knotweed is native to East Asia but has become a fast-spreading weed in temperate regions of the world. Its aggressive root system helps it establish in challenging environments, such as roadsides and in soil contaminated with heavy metals. It has even colonized volcanoes, including the iconic Mount Fuji.

7/ **Natural anti-freeze**
Polypodium vulgare
The common polypody is a hardy evergreen fern that grows on rocks and mossy branches in Europe and Asia. It is able to survive very low temperatures by increasing the concentration of sugar in its leaves. This lowers the freezing point of water and stops sharp ice crystals from forming, which could damage leaf cell walls.

8 /

9a /

10 /

11 /

• TEMPERATE FOREST •

8/ Obstacle course
Cypripedium calceolus
Native to Europe and Asia, the lady's slipper orchid is pollinated by small bees and flies, which are lured by its showy, scented flowers. When an insect lands on the pouch-shaped central petal of a flower – the "slipper" – it loses its footing on the slick surface and falls inside. The only escape is through two small holes at the back of the pouch. These gaps force the insect to rub against a sticky stigma, depositing any pollen it carries from another flower. Then it travels past the anthers, which cover its back with pollen, ready to be deposited when it falls for the next flower trap.

9/ Prolific parasite
Viscum album
Mistletoe is a widespread evergreen shrub in European lowland forests. It is hemiparasitic – living as a parasite on trees, but also performing photosynthesis. It prefers open habitats with plenty of light and is highly adaptable, growing on 452 identified host species. While it can affect the growth and vitality of hosts, it provides food and shelter for birds and insects, and also creates nutrient-rich leaf litter.

10/ Fruitful climber
Actinidia chinensis
This fast-growing vine is found in dense oak forests and mountainous ravines in southern China. Its heart-shaped leaves can grow to over 25 cm (10 in) across, and it has large, slightly scented white or cream-coloured flowers, followed by sweet, fleshy golden fruits. These attract birds, bats, and small mammals, who consume them and disperse the seeds in their droppings. The fruits have also enticed humans – this species is the common ancestor of many commercial varieties of kiwi fruit.

11/ Mimicking mushrooms
Asarum caudatum
Called tailed snakeroot for the tail-like lobes on its flowers, this plant is a creeping perennial native to moist forest floors in North America's Pacific Northwest. Its reddish brown cup-shaped flowers, which mimic the appearance and can smell of rotting fungi, are attractive to female fungus gnats looking for a site to lay their eggs. When a gnat enters a flower, its back brushes anthers bearing pollen. It transfers this pollen to the next female flower it visits, achieving pollination.

12/ Central heating
Arum maculatum
This distinctive perennial is found in shady woodland across Europe, where it flowers early, in April. Its large, pale green sheath houses a clublike inflorescence, whose tip burns stored starches to generate a remarkable amount of heat – up to 15°C (27°F) more than the surrounding air – to attract pollinators.

Sticky berries containing seeds are ingested by visiting birds or stick to their beaks

Rootlike sinkers penetrate the bark of the host tree

9b /

Parasitic grip
Mistletoe sends rootlike structures called sinkers into a host branch to extract water and nutrients.

12 /

Purplish-brown tip of the inflorescence (spadix) produces a carrionlike odour

13/ Shady character
Asimina triloba
Found in forests in eastern North America, the pawpaw is the most temperate member of the tropical family Annonaceae. Thriving in shade, this understorey tree has shallow, spreading, underground stems that sprout to form thickets of clones. These colonies create dense shade, depriving other plants of light. This method of domination means pawpaw does not have to rely on reproducing via its fruits and seeds, which are adapted to dispersal by now-extinct animals such as mastodons and giant sloths.

14/ Brilliant bracts
Davidia involucrata
This medium-sized tree is native to the humid mountain forests of central and southwestern China. It is known as the handkerchief tree for the large white bracts, resembling fluttering handkerchiefs, that surround its small reddish purple flower heads. Each flower head contains one female flower and many pollen-producing males. The bracts serve a dual purpose – to attract pollinators to the flowers hidden within their folds, and to protect the blooms from rain.

15/ Spring herald
Hyacinthoides non-scripta
The common bluebell is native to deciduous woodlands of western Europe, covering forest floors before trees produce their leaves in spring. This early display allows it to complete its growth and flowering before leaves create dense shade. It also ensures its blooms are a first pit-stop for pollinators, supplying nectar and pollen when few other flowers are available.

16/ Family tree
Populus tremuloides
The quaking aspen is the most widely distributed tree in North America. Although it is able to reproduce sexually, via seeds, it more usually spreads by producing clones, in the form of shoots and suckers, from its lateral roots. In this way, a single plant can create an entire colony of genetically identical offspring. Because all the trees in an aspen colony are considered part of the same plant, the Pando clone in Utah, US, which covers 106 acres (43 hectares), is often described as the world's heaviest and oldest living organism.

17/ Bitter and twisted
Humulus lupulus
Wild hop is a climbing plant of hedgerows and wooded river valleys in Europe, North America, and western Asia. Unlike vines, which use tendrils or suckers, hop twists around supports using backward-pointing hairs to help it grip. Its female fruiting cones are also specially adapted to aid survival. They have glands that produce lupilin, a bitter substance that deters herbivores, and has antibacterial and antifungal properties – it also gives beer its unique flavour.

18/ Forest phantom
Monotropa uniflora
The enigmatic ghost plant lends a spectral presence to shady forests in North America, East Asia, and northern South America. It is completely white because it lives as a parasite, extracting energy from fungi rather than photosynthesizing, avoiding the need for the green pigment chlorophyll. Once a year, after rainfall, it produces pipe-shaped flower stems clothed with tiny, translucent, scalelike leaves. The stems last just one to two weeks – a survival strategy to avoid devoting costly resources to long-lasting above-ground structures.

19/ Dazzling display
Prunus speciosa
Native to coastal and mountain forests in Japan, the Oshima cherry is renowned for its abundant white blossoms with gold stamens. Its profuse floral display increases opportunities for cross-pollination with nearby cherry species – a strategy useful for creating a genetically diverse population that can withstand a variety of conditions. This talent for hybridization has also been exploited commercially – many cultivars have been bred from the Oshima cherry.

Pawpaw petals turn maroon when mature

13 /

• TEMPERATE FOREST •

14 /
15 /
16 /
17 /
18 /
19 /

Bay laurel
Laurus nobilis

Common in the Mediterranean region, the bay laurel is a widespread remnant of ancient laurel forests. According to Greek mythology, the nymph Daphne was transformed into a laurel tree to save her from the unwanted advances of the god Apollo. Still infatuated with Daphne, Apollo then fashioned wreaths of laurel to console himself.

In Ancient Greece, bay laurel was used to make wreaths to celebrate poets as well as heroes, as a symbol of merit, victory, and honour, especially in events associated with Apollo. The slightly narcotic laurel berries were said to be chewed by the high priestess of the Temple of Apollo at Delphi to help her enter a trance and provide prophecies.

The Romans used laurels to commemorate victory and as emblems of prosperity and good health, and the plant was later appropriated by Roman emperors for protection from conspiracies and accidents. Some Romans believed that the plant could ward off lightning. In seventeenth-century Europe, it was believed that laurel had a protective nature that could ward against disasters.

To this day, a laurel wreath is linked to high academic achievement, and as such, the word laurel is found in phrases such as "baccalaureate" and "poet laureate".

Inspiring art
Laurel wreaths are a recurring feature in European art. They were used by Art Nouveau artists, such as Alphonse Mucha, who adorned his female muses with floral motifs and used laurel wreaths as symbols of poetic inspiration.

Evergreen leaves contain essential oils and are highly flammable when dry

• FORESTS •

"The laurel is especially consecrated to triumphs… and guards the portals of our emperors."

Pliny the Elder, *Natural History*, c. 77–79

Bay leaves are used in cooking for their warm, aromatic flavour

The bay laurel has small flowers, yellow–green in colour, that appear in spring and attract bees and butterflies

OAKS AND ALLIES

Encompassing oaks, beeches, and chestnuts, the beech family (Fagaceae) is found mostly in the Northern Hemisphere. It occupies a range of diverse habitats, from Mediterranean to tropical forests.

Fagaceae originated in Eurasia and spread to North America and – to a very limited extent – South America. Most species in the family bear separate male and female flowers on the same tree, with the male flowers usually held in clusters called catkins. The fruits are nuts that are either fully or partially enclosed in a cuplike structure called a cupule or husk, which may be scaly or spiky.

HISTORIC MIGRATION
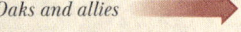
Oaks and allies

REPRESENTATIVE LIVING SPECIES

- Castanea sativa
- Fagus grandifolia
- Trigonobalanus excelsa
- Quercus acuta
- Quercus humboldtii
- Quercus robur

Quercus acuta
The Japanese evergreen oak is a small tree that belongs to the Asian ring-cupped oaks. Each cupule is covered with scales arranged in distinct rings that enclose about half the acorn.

Lance-shaped leaves cluster at the end of branches

❻ North American oaks return to Europe during the Eocene through the North Atlantic land bridge and diversify further by hybridizing with native species.

❷ Different oak lineages are established; southern European oaks spread and diversify from their Eurasian ancestors.

❸ The Bering Sea land bridge allows oaks and beeches to move from Eurasia to North America.

❺ Mountain building in Mexico isolates oak populations, leading to high diversification. The isthmus between Panama and Colombia prevents most species from spreading into South America, with one exception.

❹ Further spread and diversification takes place between east and west North America. The western species undergo a strong diversification linked with a Mediterranean climate.

Each multi-valved cupule contains several three-sided nuts

Trigonobalanus excelsa
The Colombian black oak occurs in South America, and is one of three species in this tropical genus – the other two are in Southeast Asia. This disjunct distribution is due to long-distance dispersal.

FORESTS

Quercus humboldtii
The Andean oak is an adaptable tree that adjusts to the climatic conditions at different altitudes in the Andes. Characteristics such as leaf area and wood density often vary in populations found at different elevations.

1 Oaks and their relatives originate in the Northern Hemisphere around 56 million years ago during the late Cretaceous, and expand east and west during the cooler Paleocene.

Borne on long stalks, acorns consist of a nut nestled inside a cup (cupule)

7 A Himalayan uplift about 33 million years ago creates further diversification of Southeast Asian oaks and promotes their spread into tropical forests.

Quercus robur
The pedunculate oak is the dominant tree of temperate European forests. It plays an important ecological role by providing food, shelter, and an abundance of microhabitats for a diverse range of plants and animals.

Toothed leaf margins

Spiky husks contain 1–3 chestnuts

Castanea sativa
Sweet chestnuts are long-lived deciduous trees with twisted trunks. Native to southern Europe, West Asia, and North Africa, they are grown worldwide for their edible fruits.

Fagus grandifolia
The only beech native to North America, the American beech produces a triangular, edible nut enclosed in a spiny, four-lobed husk that splits open when dry.

INVASIVE SPECIES

When non-native, or alien, species are introduced to a new environment the consequences are often unpredictable. The Spanish bluebell (*Hyacinthoides hispanica*), which first arrived in the British Isles as an ornamental garden plant at the end of the seventeenth century, is a well-known example. Over time, its presence began to threaten the UK's native species – common bluebell – *Hyacinthoides non-scripta* – as it progressively invaded broadleaf woodlands, and eagerly created new hybrids. By 2003, one in six forests in Britain contained either the larger and more vigorous Spanish bluebell or hybrids of the two species.

The danger to the common bluebell was twofold. Firstly, the more vigorous Spanish bluebell threatened to outcompete the native species for resources. Secondly, hybridization between the two species altered the genetic make up of the common bluebell and diluted its unique characteristics. Fertile hybrids spread ferociously in gardens, and it was feared that the entire population of common bluebells might become hybridized or even completely replaced by its Spanish relative.

Spanish bluebells are now so firmly established in the UK that it is almost impossible to eradicate them. Conservation efforts have focused on ensuring that horticultural suppliers provide accurately labelled bulbs, and discouraging gardeners from planting Spanish bluebells or hybrids near native populations of the common bluebell, such as those found in ancient woodlands. Interestingly, a study from 2019 led by the Royal Botanic Garden Edinburgh allowed a mixture of the two species to fertilize each other naturally. The results demonstrated that the common bluebell was more resilient than previously thought. Not only is it greater in number than the Spanish bluebell in the UK, the common bluebell also proved to be more fertile, and its seeds had a higher germination rate, giving grounds for hope that this woodland treasure will revive and thrive for many years to come.

Spring action
A carpet of bluebells covers the ground of Carstramon Wood in south-west Scotland in May. Although they take time to establish and spread, common bluebells are useful biological indicators of the health of an ancient woodland.

· FORESTS ·

PLANTS OF THE
BOREAL FOREST

Boreal forests, also known as taiga, represent the largest sub-biome on Earth. They are found in the high-latitude regions of North America, Europe, and Asia, and consist mainly of coniferous trees with the ability to endure a subarctic climate characterized by long, cold, dry winters and short, moist summers. Organic matter does not break down easily in the cold conditions, so the soil is generally acidic and low in nutrients.

1/ Easygoing pine
Pinus sylvestris
Although this pine is part of the boreal forest, it is equally happy in warmer climes, and its vast range – stretching from western Europe to eastern Siberia, and from Anatolia and Caucasia to the Arctic Circle – reflects this. Its widespread success is largely due to its ability to thrive in poor, sandy soils, where other trees have difficulty establishing.

2/ Protective shield
Silene acaulis
Found in areas where boreal forests meet high tundra, this Arctic campion flourishes in well-drained, exposed conditions. It forms low, dense cushions that progressively expand. This compact growth habit traps heat and provides protection from icy winds, creating a microclimate that can even sustain other, less hardy organisms.

3/ Good and bad
Caragana arborescens
Native to Siberia and East Asia, the Siberian pea tree contributes to its environment in several ways. Its nitrogen-fixing roots improve the soil and prevent erosion, and its dense growth functions as a natural windbreak. However, these benevolent traits are overshadowed by its aggressive spread in regions such as North America, where it now threatens native flora.

4/ Trapping heat
Papaver radicatum
Arctic poppies enjoy the open rocky habitats that occur in all circumboreal regions of the world. Black hairs on their stems and leaves absorb heat to provide insulation, and an ability to follow the sun allows them to maximize exposure to light. Cup-shaped flowers – also covered in black hairs – trap heat, which creates a warm refuge for pollinators.

• BOREAL FOREST •

Mature five-valved capsules split open to release numerous seeds

5 /

5/ A bog basic
Andromeda polifolia
A shrub in the heath family, bog rosemary grows in acidic bogs in northern polar regions. This evergreen requires plenty of moisture and has leathery leaves that reduce water loss. The pink or pinkish white, urn-shaped flowers are able to self-pollinate, but the plant can also reproduce via rhizomes – an ability that is useful in the aftermath of fires.

6 /

8a /

7 /

9 /

10 /

• BOREAL FOREST •

Pair of parasites
Tapping into the underground network of fungal hyphae (threadlike filaments), the ghost orchid relies on a symbiotic relationship between trees and fungi – in which sugars are exchanged for water and minerals – to supply its nutrition.

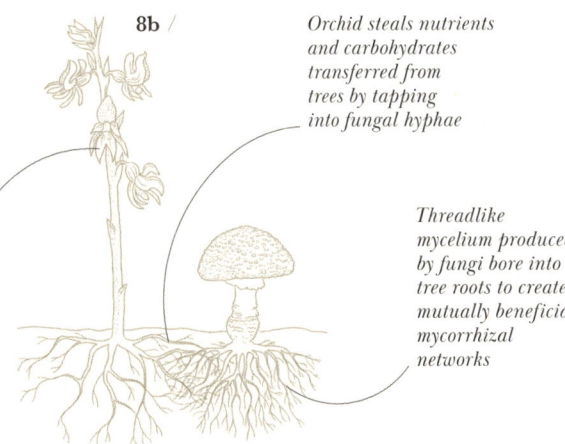

8b /

Orchid steals nutrients and carbohydrates transferred from trees by tapping into fungal hyphae

Powered by stolen energy, the orchid flowers and sets seed after self-pollination or visits from insects such as bumblebees

Threadlike mycelium produced by fungi bore into tree roots to create mutually beneficial mycorrhizal networks

6/ Speedy settler
Chamaenerion angustifolium
Distributed widely across high latitude areas, this willowherb is a pioneer species that rapidly colonizes open spaces without competition, such as forest clearings or after fires. Each plant produces a vast amount of seeds with silky hairs. The seeds swiftly establish themselves when they land on disturbed ground, and then extend their territory using a large network of shallow underground roots.

7/ Fatal trap
Sarracenia purpurea
An inhabitant of eastern North America, this carnivorous plant has specialized leaves akin to pitchers. Covered in smooth hairs, the waxy lip of the leaf exudes a nectar that attracts insects, spiders, and newts. On arrival, these prey slip down into the base – which is filled with water and digestive enzymes – where they drown, and release their nutrients to the plant. Older leaves act as long-lasting "stomachs", storing nutrients.

8/ Forest spectre
Epipogium aphyllum
Native to dense forests in northern Europe and Asia, the ghost orchid has no leaves. As it lacks the chlorophyll necessary for photosynthesis, it relies on decomposing matter and parasitizing fungi – which are themselves parasites – to provide the nutrients it needs. This epiparasitic, opportunistic existence makes it difficult to predict where the orchid might grow.

9/ Responsive leaves
Rhododendron groenlandicum
Native to Greenland and subarctic North America, Labrador tea has thick, leathery leaves with a woolly mat of hairs on the underside – features that reduce surface area and minimize water loss in its cold, windy environment. The leaves also curl inwards and droop in very low temperatures, mechanisms that reduce sun damage and protect cells by slowing thawing.

10/ Vital food source
Rubus chamaemorus
The cloudberry occurs in all Arctic regions, but it is also found in parts of Europe as a relic of the Ice Age. As male and female flowers are borne on separate plants, it requires insect pollinators to transfer pollen from one to the other, and rewards them with nectar. Its leaves are a vital food source for caterpillars of several moth species, while its raspberrylike, vitamin-rich fruits sustain diverse mammals and birds.

11/ Pollen catapult
Cornus canadensis
The creeping dogwood is a carpet-forming perennial from North America and northeast Asia, including Japan. Its tiny green flowers have an unusual mechanism to aid pollination. When a pollinator brushes a sensitive trigger on the petals, they flip open, releasing spring-loaded stamens that blast pollen into the air and onto the insect.

Showy white bracts attract pollinators to the tiny green flowers at the centre

11 /

12/ Colourful cue
Melampyrum nemorosum
This annual from European forest margins is regarded as a hemiparasite, because it draws nutrients from the roots of various hosts, even though it is capable of photosynthesis. In summer, it produces spikes of bright yellow tubular flowers on stems topped by purple bracts. This striking colour combination acts as a strong visual signal to pollinating insects.

13/ Cool character
Larix gmelinii
The East Siberian larch is the dominant tree in Northeast Asia, forming enormous forests, even on inhospitable permafrost. It is unique in being the northernmost tree in the world and also the tree most resistant against frost, surviving temperatures as low as −70°C (−94°F). Its very shallow root system helps it to easily access water and nutrients from any meltwater and from the seasonally thawing layer of permafrost.

14/ Scrappy spreader
Lycopodium annotinum
This small relative of ferns, found across all northern areas, is called a "guerrilla lycopod" for its ability to scramble over rocks and barren areas only to put down roots wherever tiny patches of soil are available. It has tough roots and shoots, forms vast mats, and is extremely efficient in mineral uptake and recycling, a necessity in the hostile boreal environment.

15/ An antifreeze system
Vaccinium uliginosum
Bog bilberry is a hardy shrub adapted to survive temperatures below −40°C (−40°F). It copes with the cold by storing sugars and proteins, which serve as antifreeze compounds, in the cells of vulnerable areas, such as buds and stems. Flexible cell membranes withstand the stress of freezing and thawing cycles, while moisture is moved to intercellular spaces to reduce damage from ice formation.

16/ Sticky endings
Drosera rotundifolia
At home in acidic habitats across the Northern Hemisphere, sundew has turned to carnivory to get the nutrients it needs. Specialized leaves with glistening hairlike tentacles tempt insects to investigate – which are quickly stuck fast and engulfed by a curling leaf. To avoid accidentally digesting pollinators, sundews produce flowers in contrasting colours that stand proud of the traps on long coiled leafless stems.

17/ Established hybrid
Sorbus sudetica
This Central European rowan is likely derived from cross-fertilization between *Sorbus aria* and *S. chamaemespilus*. Its survival as a successful hybrid after this chance event lies in its ability to produce its own seeds asexually – without the need for normal fertilization. This means that offspring are genetic clones of the parent, ensuring the *S. sudetica* lineage continues.

18/ A benefactor to the forest
Shepherdia canadensis
This small deciduous shrub occurs in North America. In boreal forests, where nitrogen necessary for plant growth and nutrition is sparse, *Shepherdia* has the ability to fix nitrogen. It does so by forming a symbiotic relationship with the bacterium *Frankia*, which infects the plant's roots to form nodules that convert atmospheric nitrogen into usable ammonia.

12 /

Blue-violet bracts attract insects

Two-lipped, tubular flowers appear on a terminal spike

• BOREAL FOREST •

13 /

16 /

17 /

14 /   15 /

18 /

A noble flower

Linnaea borealis became a favourite plant of Carl Linnaeus following his encounters with it in Lapland in 1732. He even commissioned the plant to be painted trailing around his coat of arms when he was ennobled in 1761.

Twinflower

Linnaea borealis

Stems creep along the ground forming a mat

Known as the twinflower for its bell-shaped flowers that grow in pairs, *Linnaea borealis* is a small creeping herb found in moist subarctic forests around the world. It has long been used in traditional medicine by different cultures. In 18th-century Norway it was recorded as a folk remedy for shingles, skin conditions including eczema, and as a treatment for rheumatism. In North America, First Nations people found uses for the plant as a tonic during pregnancy and as a treatment for colds, headaches, and fevers.

The plant is also strongly associated with the Swedish botanist Carl Linnaeus, who developed the system of classifying organisms by their genus and species names. Although he initially gave the twinflower another name – *Campanula serpyllifolia* – the modest plant was named after Linnaeus in 1737 by the Dutch botanist Jan Frederik Gronovius. In Sweden, the strong connection of the flower with Linnaeus continues by its recognition as the flower emblem of Småland, in southern Sweden where Linnaeus was born. The beautiful but ephemeral flower has also come to be associated with modesty and inspires the given name Linnea, which has become popular in Sweden and Norway in recent years.

> *"Insignificant, disregarded, flowering but for a brief space – from Linnaeus who resembles it."*

Carl Linnaeus, *Critica Botanica,* 1737

• FORESTS •

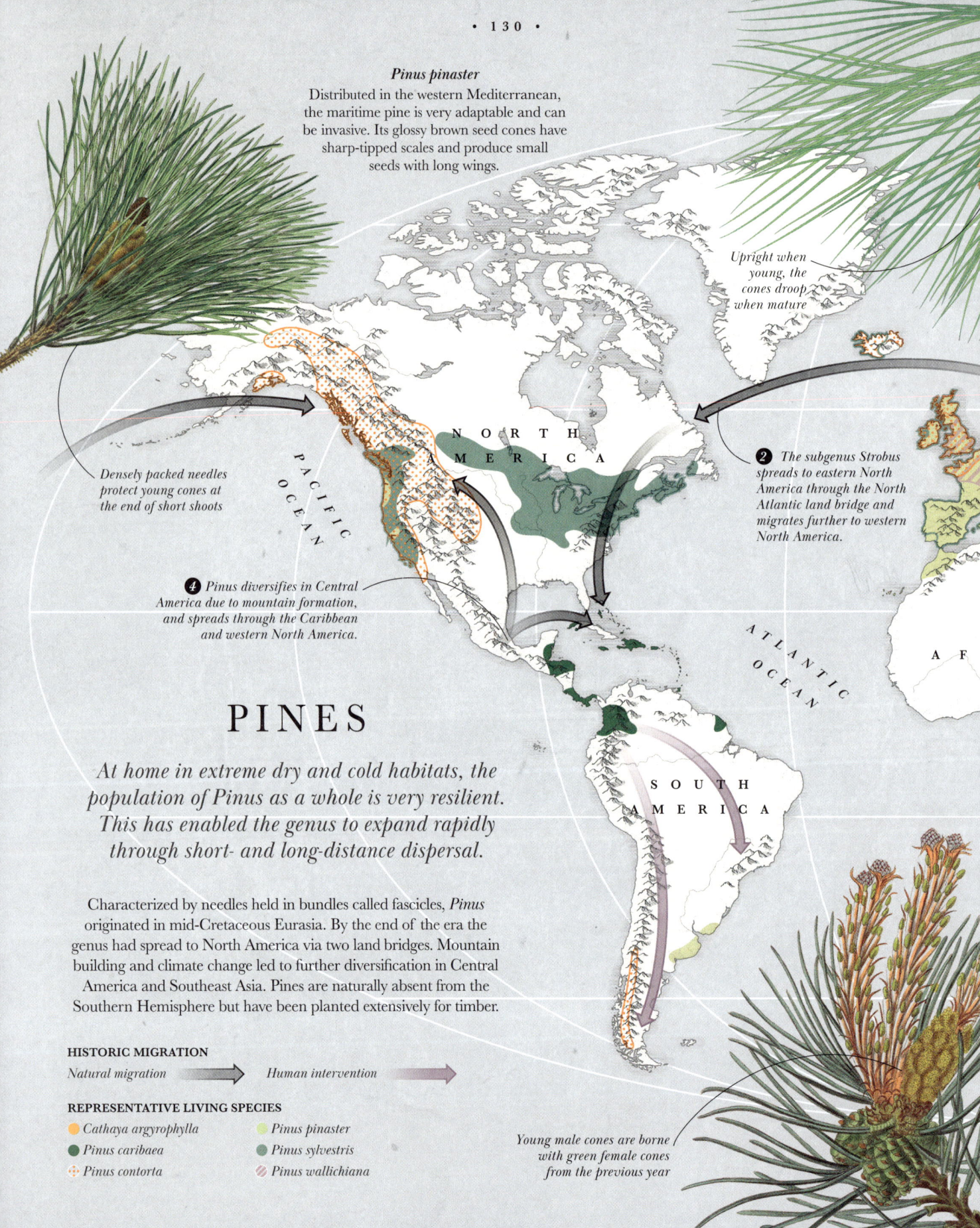

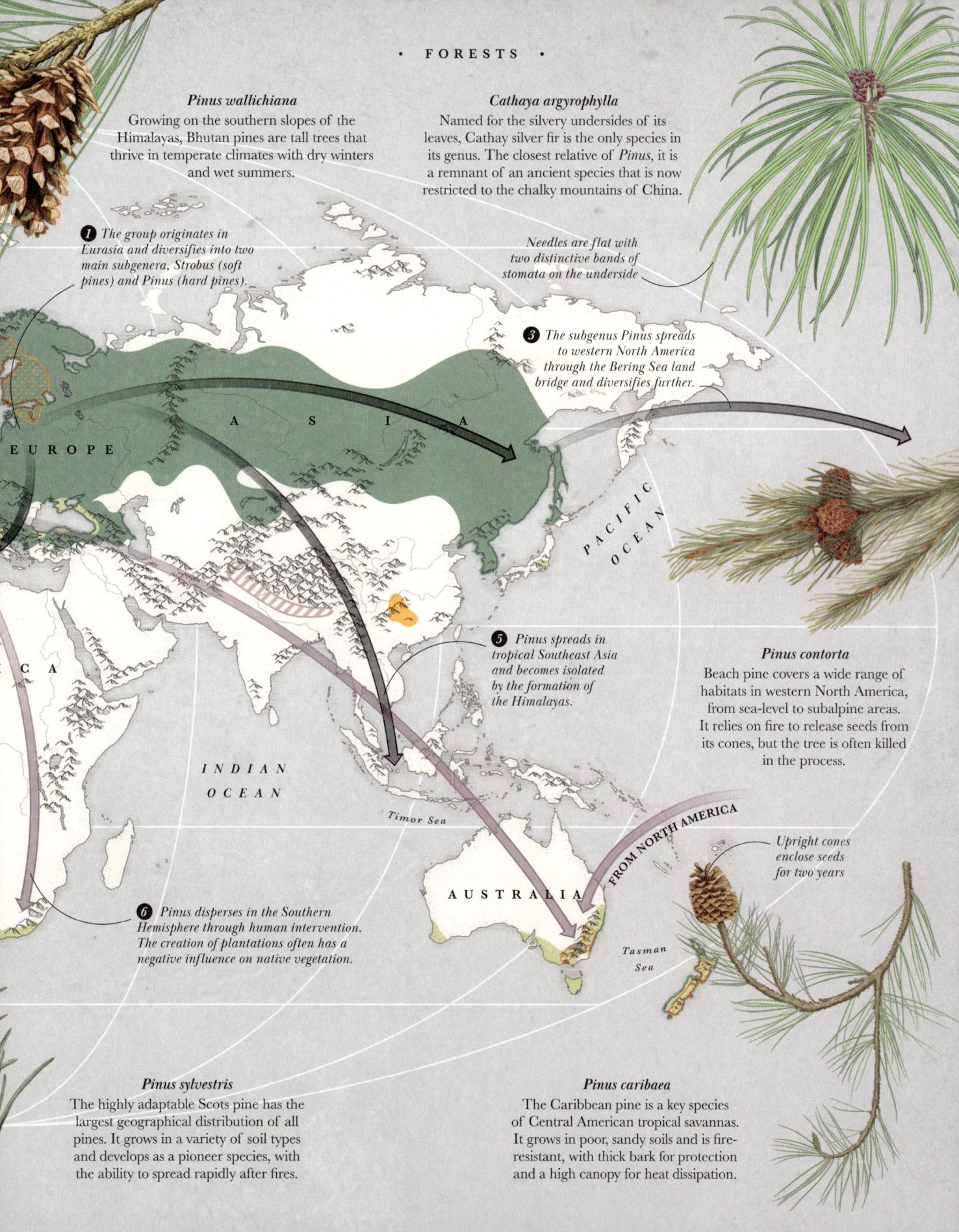

Cacao

Theobroma cacao

The genus name *Theobroma* means "food of the gods" – and nothing embodies that name more than cacao (*T. cacao*), which is used to produce chocolate-based food and drink that is highly desired all over the world. Originally from the Amazon forest and introduced to Central America – where it was associated with Mayan mythology – cacao spread to the rest of the world after the Spanish conquest in the 16th century, and it is now grown in several tropical countries, mainly in Africa and South America.

The plant produces substantial berrylike fruits that contain large seeds known as cacao beans. The base of all cacao products, the bean is fully fermented, dried, and roasted, then ground up to make a paste. The Mayans and Aztecs used this paste as a bitter drink spiced with chilli and vanilla. Viewed as a connection to the gods, this cacao drink played an important part in religious rituals, and its use was restricted to priests and the ruling aristocracy. The same drink later became popular in Europe after the Spanish sweetened it with sugar.

Modern chocolate was developed in the 19th century with the invention of a hydraulic press that separated cocoa butter from the beans, leaving a fine, powdery residue. The addition of milk and sugar to the cocoa powder resulted in the creation of milk chocolate.

"Chocolate symbolizes, as does no other food, luxury, comfort... gratification, and love."

Karl Petzke and Sara Slavin, *Chocolate: A Sweet Indulgence,* 1997

Leathery leaves are shed periodically

Cacao beans are enclosed in a white edible pulp

Ribbed fruits may be yellow, orange, red, or purple in colour

Chapter 2

GRASSLANDS
& SHRUBLANDS

Flammable landscapes

Fire occurs in all grasslands and shrublands, but varies in severity. The most intense and frequent fires are usually found in the tropics and subtropics, where grasses accumulate enough biomass to burn every 1–3 years.

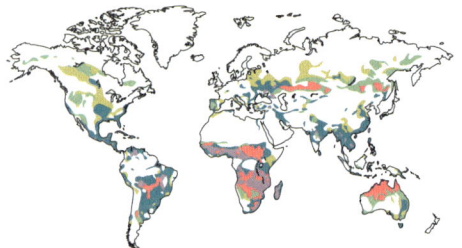

KEY
- Frequent; intense; large
- Frequent; cool; small
- Rare; intense; large
- Rare; cool; small
- Intermediate; cool; small

The Great Plains are a rare and extensive area of temperate grassland

The Cerrado is the world's most species-rich tropical grassland

GRASSLANDS AND SHRUBLANDS

About 40 per cent of Earth's land surface is covered by grassland and shrubland. These open areas are home to grasses and shrubs that can tolerate a range of threats, including those caused by seasonal drought, fire, grazing animals, or a combination of different factors.

INFINITE VARIETY

Grasslands and shrublands encompass lush, species-rich tropical savannas, arid, sparsely vegetated steppes, and cold, tundralike shrublands. Warm regions have abundant grasses with adaptations to grazing or fire, while cooler and drier regions support species adapted to nutrient-poor soils, extremes of temperature, and short growing seasons.

Seasonal shifts

Cold and temperate grasslands and shrublands experience seasonal variability in both temperature and rainfall. They are mostly in sync, with warm temperatures and high rainfall coinciding in summer. In the tropics, temperatures remain consistent but plant life experiences seasonal drought.

KEY
- Cold regions
- Temperate regions
- Tropical regions
- Average temperature

ADAPTATIONS

GROWTH AND PHOTOSYNTHESIS

Grasslands are open biomes, often with plenty of light, moving air, and a high diversity of animals. Grassland soils can be nutrient-poor and fairly dry, but areas of wetter, richer soil can harbour more vigorous species. Areas disturbed by trampling, such as migration pathways, often provide new opportunities for weedy species to settle. The open aspect of grasslands has a profound influence on both photosynthesis and the dispersal of pollen and seeds. In tropical areas, where there is plenty of sunshine and temperatures are high, some grasses have developed strategies to optimize their photosynthetic potential.

fig. 1 Cortaderia selloana; fig. 2 Poa annua; fig. 3 Agropyron cristatum; fig. 4 Briza maxima; fig. 5 Themeda triandra; fig. 6 Pennisetum villosum; fig. 7 Ophrys apifera; fig. 8 Proboscidea louisianica; fig. 9 Tragopogon pratensis

fig. 1

Growth forms

Grasses in dry, wet, and disturbed areas favour different growth forms. Tillering grasses spread over large areas, often in drier soils. Tussock-forming grasses are more characteristic of moister soils, and annual grasses are opportunists from disturbed areas.

Dense, upright tussocks persist for several years

Tough, unpalatable leaves

Tussock grass

fig. 2

Short-lived annuals send up multiple flower heads to ensure they produce plenty of seed

Annual

fig. 3

Tillers branch off the parent plant and root at nodes

Tillering grass

• GRASSLANDS AND SHRUBLANDS •

C3 and C4 photosynthesis

Like most green plants, grasses take in carbon dioxide (CO_2) and release water and oxygen in a process properly known as C3 photosynthesis. However, in regions that experience high temperatures, low water levels, and lots of light, several groups of tropical and subtropical plants, including grasses, have evolved a more efficient system – C4 photosynthesis – which concentrates the CO_2, producing better growth and using less water.

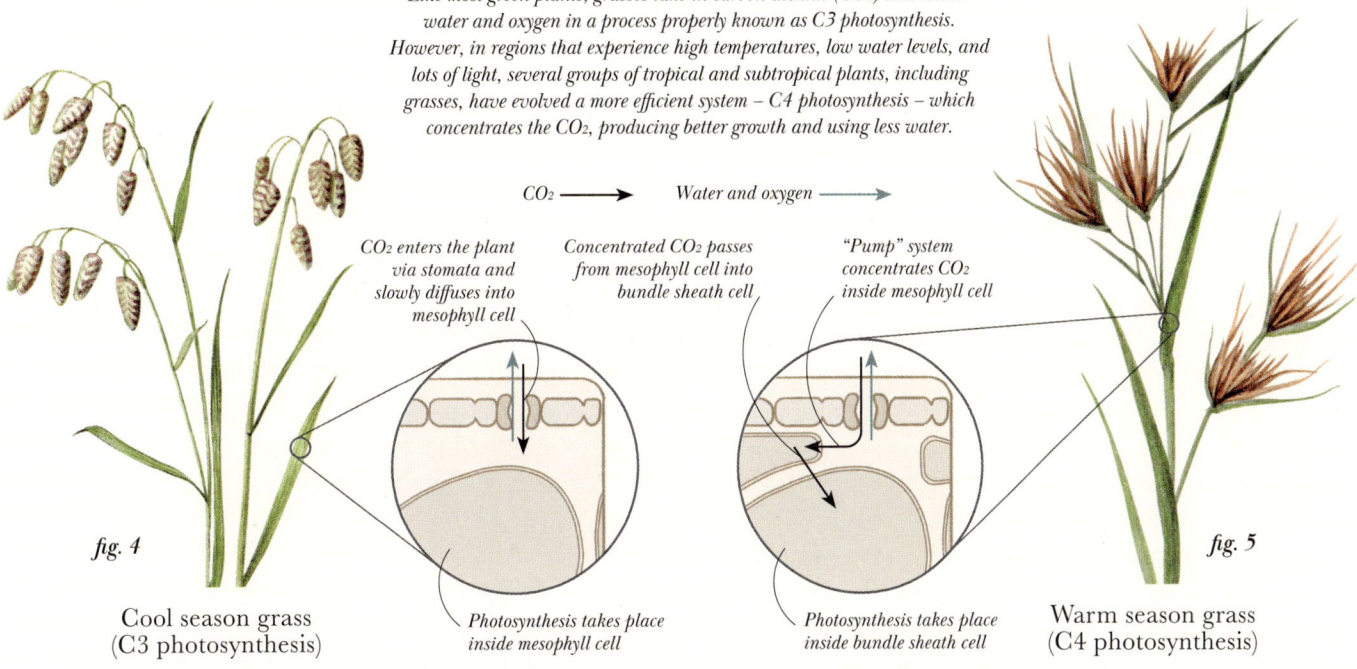

CO_2 ⟶   Water and oxygen ⟶

CO_2 enters the plant via stomata and slowly diffuses into mesophyll cell

Concentrated CO_2 passes from mesophyll cell into bundle sheath cell

"Pump" system concentrates CO_2 inside mesophyll cell

fig. 4

Cool season grass
(C3 photosynthesis)

Photosynthesis takes place inside mesophyll cell

Photosynthesis takes place inside bundle sheath cell

fig. 5

Warm season grass
(C4 photosynthesis)

Pollination strategies

Grass florets are individual flowers that do not have petals or sepals. This adaptation allows the florets to collect pollen more efficiently when it is released into the air. Other grassland plants such as orchids deploy more targeted strategies to ensure pollination.

Flower mimics a female bee to trick a male bee into "mating", but it picks up pollen instead

Male stamens hang out of the flower head

fig. 6

Wind pollination

fig. 7

Pollination by deceit

Seed dispersal strategies

Grassland plants exploit both animals and wind to ensure their fruits and seeds are dispersed. Those that are eaten may travel long distances inside migrating herd animals, while others hitch a ride on fur, feathers, or hooves.

Feathery pappus carries light fruits vast distances

Hooked fruits mature and drop, ready to catch onto animals for dispersal

fig. 8

Animal dispersal

fig. 9

Wind dispersed

ADAPTATIONS

FIRE AND PREDATORS

Grazing – and in some areas, fire – are crucial factors defining the growth form, regeneration abilities, and many other adaptations of steppe and savanna grasses. Some adaptations protect against grazers, but also do other jobs. In the Mediterranean, for example, many shrubs, especially members of the mint (*Lamiaceae*) and daisy (*Asteraceae*) families, have volatile oils that perfume hillsides on hot days. These volatile oils are thought to deter herbivores, attract pollinators, and may even help to cool the plants – like a scented botanical sweat. The dense oil vapours may also act like a cocoon of gas, preventing water from evaporating into the drier surrounding air.

fig. 1 Euphorbia stepposa; fig. 2 Senegalia senegal; fig. 3 Salvia pratensis; fig. 4 Fritillaria imperialis; fig. 5 Digitaria macroblephara; fig. 6 Vachellia karroo; fig. 7 Enterolobium gummiferum; fig. 8 Aldama grandiflora

Defensive strategies

Grassland plants have an array of defences and deterrents against grazing and browsing animals. Poisons, spines, and warning odours all help to protect leaves and stems. In dry areas where plants are a key source of water for animals, these adaptations become even more critical.

Poison — *fig. 1*

Any or all plant tissues may contain caustic or toxic substances such as milky latex

Spines — *fig. 2*

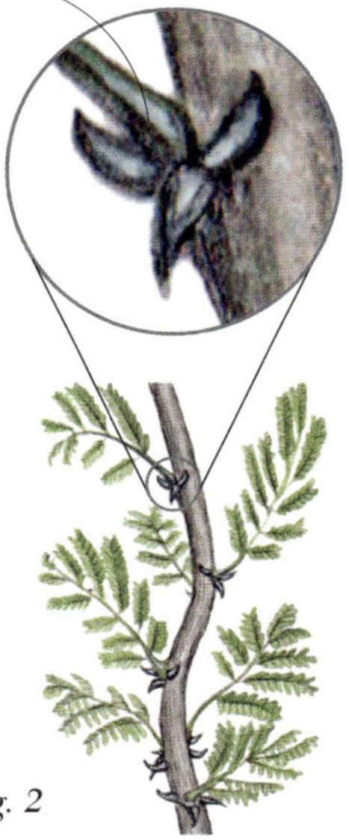

Vicious, hooked spines make browsing difficult

Volatile oils — *fig. 3*

Flowers, stems, and leaves may release volatile deterrent oils in response to threats

fig. 4

A tuft of modified leaves allows for additional photosynthesis and protects the flowers below

Flowers attract insect pollinators from grasslands and forest margins

GRASSLANDS AND SHRUBLANDS

fig. 5

Hidden growth points — Bulb stores carbohydrates and other resources that promote regrowth if crown is damaged above ground

Silica crystals in leaf — Silica granules damage the mouthparts of insect grazers

Grains of glassy silica constantly wear down the teeth of grazing animals

Surviving grazing

Bulbs and tubers that hide buds below ground, and intercalary meristems (growth tissue) of grasses sitting low to the ground allow plants to regrow when the tops are grazed off. Grass leaves also wage a war of attrition against the grazers — slowly wearing down their teeth.

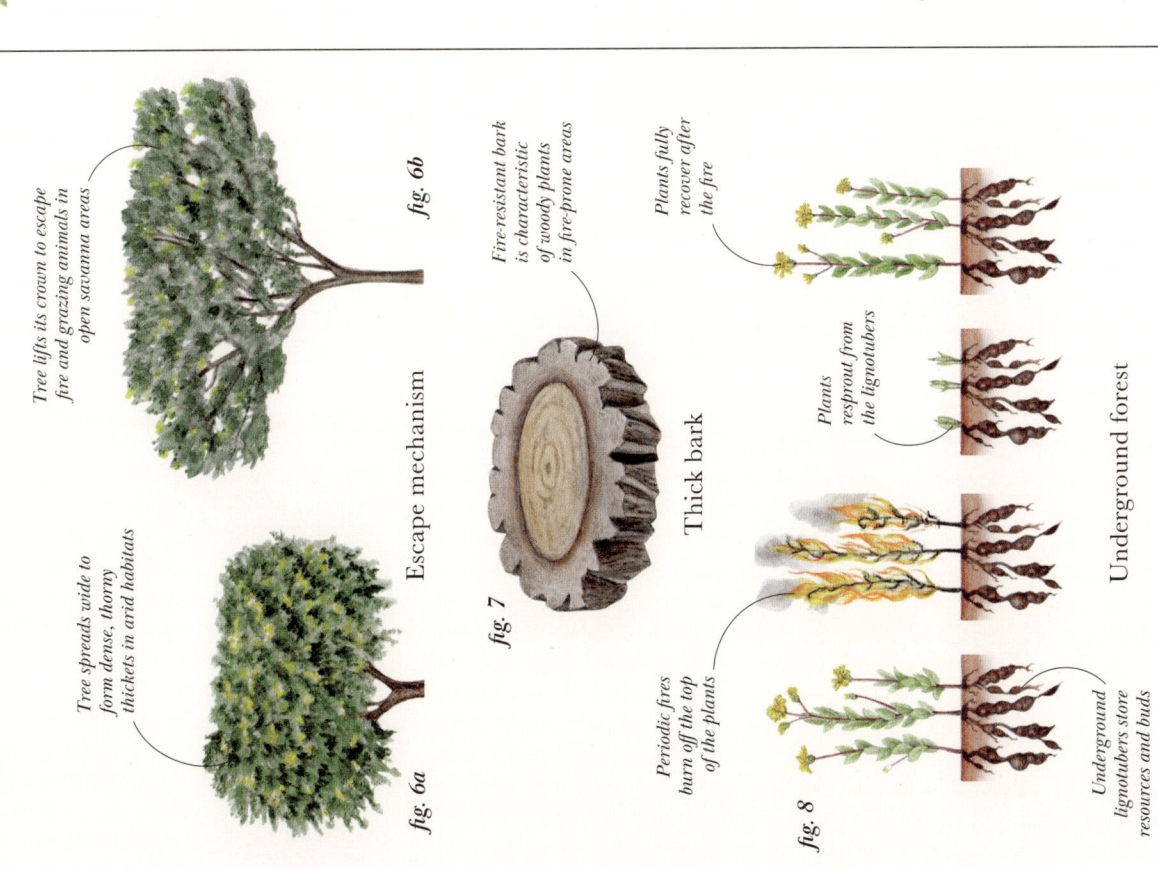

fig. 6a — Tree spreads wide to form dense, thorny thickets in arid habitats

fig. 6b — Tree lifts its crown to escape fire and grazing animals in open savanna areas

Escape mechanism

fig. 7 — Fire-resistant bark is characteristic of woody plants in fire-prone areas

Thick bark

fig. 8 — Periodic fires burn off the top of the plants; Plants resprout from the lignotubers; Plants fully recover after the fire; Underground lignotubers store resources and buds

Underground forest

Fire strategies

In fire-prone habitats, some species protect their resources by growing corky bark. Others live as "underground forests" — plants with tubers or extensive, and sometimes large, woody trunks buried deep in the soil. The above-ground parts can be burned off, but renew once the fire has passed.

PLANTS OF
COLD GRASSLANDS AND SHRUBLANDS

The frostbound landscapes of the Arctic tundra encircling the North Pole have some of the coldest conditions on the planet – and the shortest growing season. Although there are only about 60 days with sufficient warmth and light for plants to grow and reproduce, some 1,700 species have adapted to cope with low-nutrient soils, very little rainfall, and biting winds. Plants that grow on the Eurasian steppes and in the alpine tundra above the tree line on mountains throughout the world have a significantly longer growing season, but endure similarly cold and exposed conditions.

1/

2/

3/

4/

Upper lip resembles a beak

1/ Fertilizer factory
Caragana microphylla
In the cold northern steppes of China, the littleleaf peashrub forms a symbiotic relationship with rhizobia bacteria inside its root nodules in order to convert atmospheric nitrogen into a usable form. As the plant's parts die and decompose the nitrogen stored in its stems and leaves returns to the soil, benefiting nearby plants.

2/ Southern cushion
Colobanthus quitensis
The Antarctic pearlwort, is one of only two flowering plants that can survive in the harsh conditions of the West Antarctic peninsula. In order to endure some of the most extreme conditions on Earth it takes the form of a mosslike cushion. Its survival kit also includes the ability to "supercool", achieved by accumulating soluble carbohydrates such as sucrose in its cells. This lowers the freezing point of water in the cells so that it remains liquid even at very cold temperatures.

3/ Enduring networks
Carex bigelowii
The angular stems of stiff sedge stand to attention above the tree line in arctic-alpine regions. Through its extensive rhizome system, and with the help of surface-running stolons – a creeping stem that takes root at intervals along its length – the plant's tufty clumps occasionally dominate mountain communities, and serve as a valuable food source for animals such as lemmings and reindeer.

4/ Daylight robbery
Pedicularis rostratospicata
In the low-nutrient soils typical of many European alpine meadows, beaked lousewort has found a way to boost its nutrient intake. Although it can make its own food through photosynthesis, this semi-parasitic plant augments its diet by syphoning off minerals and water from its neighbours using modified root structures called haustoria that tap into its host's roots.

• COLD GRASSLANDS AND SHRUBLANDS •

Small pink or white blooms develop into edible red berries

5 /

5/ Spring breakthrough
Vaccinium vitis-idaea
In the freezing conditions of the Arctic, the lingonberry is a hardy shrub that maximizes its growing season by photosynthesizing beneath the snow. In spring, the snow layer insulates the plant, keeping temperatures above freezing, but still allows some light penetration, and traps CO_2, which allows the evergreen lingonberry to photosynthesize. By starting this process early, the plant is primed to reach maximum photosynthetic efficiency as soon as the snow melts.

6a / 8 / 9 /

7 /

COLD GRASSLANDS AND SHRUBLANDS

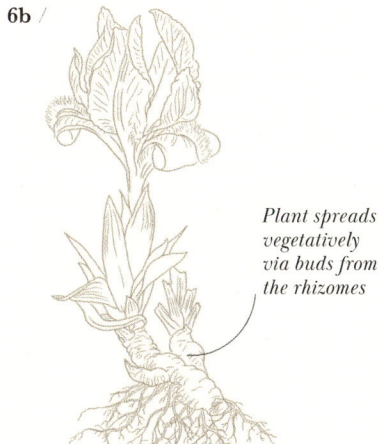

6b /
Plant spreads vegetatively via buds from the rhizomes

10 /
Pink flowers are arranged in dense, showy racemes

Natural clones
Like many rhizomatous plants, irises can produce offsets that become new plants – a form of vegetative, or asexual, reproduction. Rhizomes are underground stems that grow laterally, sprouting new shoots that develop into genetically identical clones of the parent plant.

6/ Replicating success
Iris pumila
In rocky ravines and gorges of the Eurasian steppes, this iris uses the ability to clone itself – making new plants from its rhizomes – to ensure it survives. Its populations have become fragmented due to agricultural land use. The need to adapt to slightly different environments has probably driven the genetic diversity of the populations.

7/ Persistent rarity
Pulsatilla vernalis
Covered in silky, insulating hairs, the low-growing spring pasque flower, also known as lady of the snows, clings on despite the threats posed by habitat destruction and low seed dispersal. Once found growing across the mountain grasslands and heathlands of Europe, it is now confined to nine remnant populations. A reintroduction project successfully boosted its numbers fourfold by growing wild-collected seeds and reintroducing young plants.

8/ Mighty microbes
Deschampsia antarctica
One of only two species of flowering plants found on the Western Antarctic peninsula, Antarctic hair grass is adapted to drought, flooding, ultraviolet radiation, cold-freeze-thaw cycles, and varying nutrient concentration levels. Some of its success may be attributed to a stable root-associated microbiome – a diverse community of microbes that help the plant function – which is passed between sequential generations through vegetative reproduction.

9/ Climate change hero
Phleum alpinum
Alpine Timothy grows on rocky mountainous outcrops in both the northern and southern hemispheres, as well as on the inhospitable Antarctic island of South Georgia. Studies show it is tolerant of high-wind conditions and high CO_2 levels. This means that it is likely to thrive under climate change conditions of increasing atmospheric CO_2 and more extreme weather.

10/ Ice Age survivor
Astragalus onobrychis
Equally at home in the meadows, steppes, and conifer forests of central Europe, evidence suggests that sainfoin milk vetch diversified after the last glacial period. As the ice receded, the new vegetation that sprang up competed for resources and created barriers that broke up pre-existing plant populations. This species responded by adapting differently in different places to suit the disparate soils and climates it enjoys today.

Flower colours range from a rare sky blue to violet

13/ Avoiding competition
Schivereckia podolica
A cold-adapted survivor of multiple glacial–interglacial cycles, the cushion plant avoids competition by growing in some of the most inhospitable conditions found across the Eurasian steppe. While in the past its ecological niche may have been broader, in the warmer interglacial conditions that are less favourable for it, this plant is limited to extreme rocky outcrops at high altitude.

14/ Branching out
Betula nana
In the tough conditions of the Arctic tundra, dwarf birch uses a clever strategy to see it through long winters and maximize growth in brief summers. When resources are scarce, it maintains many short shoots with only a few leaves; when resources are plentiful, these stems power into growth and produce more shoots and leaves. This ability to branch rapidly creates dense mats of low-growing foliage that capture heat and insulate the roots and stems from cold winds and freezing temperatures; it also allows the plant to renew itself by creating new offshoots that root as clones.

15/ Insulated parachutes
Eriophorum vaginatum
Supremely tolerant of fluctuating extremes in the Arctic tundra, the tufty seedheads of hare's tail cottongrass appear well prepared for chilly weather. The silky white fibres serve a dual purpose: initially they provide insulation for the plant's reproductive organs, and then they convert into parachutes that catch the wind to disperse their seeds.

11/ Isolated retreat
Meconopsis simplicifolia
Native to high alpine meadows in the Himalayas, this plant needs cool temperatures, sufficient moisture, and low competition from other plants to thrive. However, rapidly warming temperatures are encouraging plants from lower slopes to encroach on its habitat, driving this poppy further north and into higher altitudes – and potentially towards extinction.

12/ Peaty pedestal
Poa flabellata
Tussac grass is the most dominant species in the extreme environments of the sub-Antarctic islands, where it creates widespread single-species plains. These extensive tussac grass communities have existed virtually unchanged for 12,500 years, growing on pedestals more than 3 m (10 ft) tall made of their own decaying leaves and roots, and forming carbon-rich peat deposits that can be more than 13 m (43 ft) deep.

16/ Special relationship
Gymnadenia miniata
Found in the chalky grasslands of the Eastern Alps and the Carpathian mountains, the red vanilla orchid produces minute, dustlike seeds. Because they lack an endosperm (the part of a seed that stores nutrients), they can only grow into a seedling if a mycorrhizal fungus supplies them with food. This relationship has sustained this species since before the last Ice Age.

• COLD GRASSLANDS AND SHRUBLANDS •

12 /

13 /

14 /

15 /

16 /

Leaves grow close to the stem before curling away and tapering to a point

Urn-shaped flowers produced in large masses

Gaylussacia brasiliensis
Found in the shrublands and grasslands of Brazil, this plant flowers continuously rather than seasonally to outcompete other hummingbird-pollinated plants.

4 Around 16 million years ago, Gaylussacia disperses into South America – especially Brazil – where it diversifies into many more species.

HEATHERS

Originating in low-nutrient, acidic habitats in the Northern Hemisphere, the heather family (Ericaceae) is widespread, with about 4,500 species, including the large genera of Rhododendron and Erica.

Between 90–65 million years ago, when the continents were more closely connected, Ericaceae began to diversify across North America and Eurasia, eventually establishing in the tropics and subtropics 40–23 million years ago. The flowers evolved significantly as they came into contact with new pollinators, adapting from wind and bee pollinators to partners such as hummingbirds, moths, and mammals.

Creamy white flowers are sweetly scented

Woollsia pungens
Moth-pollinated snow wreath is restricted to eastern Australia where it thrives in fire-prone heath and open woodland by re-establishing itself from seeds stored in the soil.

HISTORIC MIGRATION
Gaylussacia → Erica → Rhododendron

REPRESENTATIVE LIVING SPECIES
- Calluna vulgaris
- Erica coccinea
- Gaylussacia brasiliensis
- Rhododendron ferrugineum
- Richea sprengelioides
- Woollsia pungens

PLANTS OF
TEMPERATE GRASSLANDS, SAVANNA, AND SHRUBLANDS

Seasonal rainfall and fluctuating temperatures characterize the world's temperate grasslands, shrublands, and savannas. Although the brightly coloured shrubs of the South African fynbos may look very different to the short grasses of the Eurasian steppe, both are sustained by cold, wet winters and hot, dry summers. Despite the nutrient-poor conditions of the fynbos, it has rich plant biodiversity, supporting adaptations such as tough, small, and long-lived leaves. In contrast, the nutrient-rich pampas of South America and great plains of North America are iconic for their wide expanses of fast-growing grasses and wildflowers.

1 /

1 / Royal beauty
Fritillaria imperialis
Across Iran and parts of West and Central Asia, the crown imperial lily forms beautiful bell-shaped blossoms. Crowned by a whorl of leaves, its nectar-rich flower clusters attract various pollinators including birds and bees. It emits a strong odour that repels pests, such as moles and squirrels, as well as browsing animals like deer. Dormant in summer, its cold-hardy bulb grows roots in winter to sprout in spring.

Flowers bloom in a variety of colours in cultivation, but are usually orange-red in the wild

• TEMPERATE GRASSLANDS, SAVANNA, AND SHRUBLANDS •

2/ Prickly pair
Eryngium yuccifolium

In the tallgrass prairies of North America, the distinctive round heads of rattlesnake master stand out amid the grass. These flowerheads – comprising tiny flowers surrounded by pointed bracts – together with the spiny margins of its long, thin, basal leaves, mount a prickly defense against deer and other browsers. A deep taproot anchors the plant firmly in the ground, allowing it to access and store water during periods of drought. The plant is able to re-establish itself in the aftermath of fire via its seedlings, which take advantage of the newly open prairie.

2/

3/ Honeyed tones
Phlomoides tuberosa

Jerusalem sage's striking whorls of pink-to-purple flowers are a common sight across the Eurasian meadow-steppe. Known as a honey plant, its flowers produce large quantities of nectar and pollen to attract insect pollinators such as bees and butterflies. It has fleshy, tuberous roots that store starch and water to tide it over during periods of dormancy and drought. Bitter-tasting chemicals in its leaves deter hungry herbivores.

3/

4/

4/ Clonal survivor
Leymus chinensis

In the semi-arid steppes of northern China, false wheatgrass has found a way to thrive. It reproduces mainly by creating clones in the form of offshoots from its rhizomes. Although it produces a few seeds, most of them are not viable or fail to germinate. Nevertheless, *Leymus chinensis* remains one of the most important native perennial forage species on the steppe.

5/ Heathland hero
Calluna vulgaris

Heather thrives in the acidic, low-nutrient soils of heathlands. In this unique habitat, leaf litter and dead stems take a long time to decompose, locking carbon into the soil rather than releasing it into the atmosphere. By contributing to this slowly decomposing organic matter, heather helps store carbon in the soil, and provides a natural defence against climate change.

Small scalelike leaves crowd on stem to reduce water loss

5/

6 /

8 /

7 /

9a /

• TEMPERATE GRASSLANDS, SAVANNA, AND SHRUBLANDS •

Special relationship
Oil-collecting rediviva bees are guided by olfactory and visual cues to the grand mask orchid. They collect fatty oil from its flowers to feed their larvae.

Twin petals sacs contain verrucae, or sacs, that secrete oil to attract pollinators

Anther sac contains pollen, which is dusted onto the hairy body of a visiting bee

9b /

6/ Restoration pioneer
Rudbeckia hirta
Named for its flowers, which have a dark, eye-like central disc surrounded by bright yellow petals, black-eyed Susan is a common North American prairie plant. In these wide plains, it is a pioneer species, able to quickly colonize new areas. Its ability to resist extreme heat and drought, to survive in poor soils, and to successfully self-seed, has made it a cornerstone plant in prairie restoration projects.

7/ Fragrant solidarity
Artemisia tridentata
The fuzzy silvery-grey leaves of big sagebrush can be found right across the sagebrush steppe of North America. When threatened by herbivory, a leaf will release offputting fragrant chemical compounds which not only alert the other leaves on the threatened individual to develop resistance, but also warn neighbouring sagebrush plants to do the same, in a process called "volatile communication".

8/ Floristic explosion
Berzelia stokoei
Although Bruniaceae is an ancient family of shrubs that originated between 60–100 million years ago, the formation of a Mediterranean climate in the Cape Floristic region of South Africa around 10 million years ago prompted a surge of evolution that resulted in many new and closely related species. Among them is Berzelia stokoei, which displays the typical family feature of small, hardy, overlapping leaves to resist water loss in the hot summers. It produces a profusion of tiny red flowers on round flowerheads for pollination by beetles and sunbirds.

9/ Oily gifts
Huttonaea grandiflora
The grand mask orchid is only found in the high grasslands of the Drakensberg mountains in South Africa. Its fringed petals are blotched with maroon speckles that act as landing strips for the female bees that visit this orchid. These endemic bees are attracted to the special oil-secreting sacs that the orchid hides in its flowers as a lure for pollinators.

10/ Flower power
Phyteuma orbiculare
This pretty perennial, known as round-headed rampion, thrives in chalk grasslands from southern England to Greece. Although its spherical purple flower heads look like individual blooms, each globe contains multiple flowers that are densely packed together. By aggregating multiple flowers together the plant increases the likelihood of pollination and dispersal, and optimizes its chances of reproductive success.

Florets curve inwards

10 /

11/ Scented deception
Ophrys sphegodes
Ancient chalk and limestone grasslands support one of the UK's rarest plants, the early spider orchid. Its flower produces a scent which, while not detectable to humans, attracts male andrena bees by imitating pheromones released by the female bee. This highly specialized pollinator interaction is threatened by warming temperatures that are shifting orchid flowering times to the extent that they are no longer completely in sync with the emergence of the male bees.

12/ Autumn splendour
Spiranthes spiralis
Commonly called autumn lady's-tresses, this orchid's Latin name derives from the tight spiral arrangement of its delicate flowers. Unlike most European orchids that bloom in spring or summer, this grassland species flowers in autumn, and only in certain years. Its leaves appear around the same time and persist through winter before dying in spring to complete the plant's "reversed" life cycle.

Flower mimics the female of the pollinating bee species

13/ Steppe pioneer
Krascheninnikovia ceratoides
A perennial shrub that characterizes the vast Eurasian steppes, Pamirian winterfat spread in tandem with the emergence and expansion of this biome. It was probably one of the first plants to inhabit the steppe, evolving special features to survive the challenging conditions in this habitat. Its leaves, stems, and fruit are covered in fine white hairs that give it a whitish appearance, and may help the plant to reduce water loss in extreme conditions.

14/ Friendly fire
Schizachyrium scoparium
Little bluestem spans the grasslands of North America, where it is found in prairies, pine, and oak savannas. It is considered an important grass as it plays a vital role in the ecosystem, providing food and shelter to a range of birds, including sparrows and other songbirds. It is specially adapted to benefit from fire, a common occurrence in grasslands, and its aboveground regrowth increases after fire occurs. Fires burn off dead plant material, increasing the availability of light, raising the soil temperature, and releasing nutrients including nitrogen – all of which enable it to flourish.

15/ Hot release
Protea spp.
Fires occur regularly in the shrubby fynbos region of South Africa. To survive and thrive under these conditions, the common sugarbush has evolved reproductive adaptations that rely on fire. After flowering, the dry bracts of the plant enclose the seeds, and retain them in a conelike structure until the heat of a fire stimulates their release.

• TEMPERATE GRASSLANDS, SAVANNA, AND SHRUBLANDS •

Yarrow

Achillea millefolium

An unassuming but resilient member of the daisy family found in temperate grasslands around the world, yarrow's cultural significance far outweighs its humble status. It has been found on Neanderthal remains, which suggests that the plant has been used medicinally or ritualistically for millennia.

In northern European traditions, yarrow has a long history as a herbal remedy – the Anglo Saxon medical text *Bald's Leechbook* lists its use in a "quieting drink" and it is still used today as a mild sedative. The Roman philosopher Apuleius recommended the plant for treating boils or lumps on the head, and the finely dissected leaves were widely used in Europe to stop bleeding. According to legend, the Greek hero Achilles learned of these blood-stanching properties from the centaur physician Chiron, and it is from Achilles that yarrow gets its scientific name.

The plant was also believed to carry magical properties and was used as a charm to protect against bad luck or offer prophetic insights. A Gaelic tradition claimed that a maiden who cut nine stalks of yarrow and laid them under her pillow would surely dream of her husband to be.

"I will pluck the yarrow fair,
That more benign shall be my face,
That more warm shall be my lips."

Alexander Carmichael, *Carmina Gadelica*, vol. 2, 1900

Daisylike flower heads attract pollinators such as honeybees

GRAZING

Grazing is a natural part of many ecosystems, but too much of it can be devastating. Overgrazing strips the land of vegetation, leaving bare ground prone to erosion and nutrient loss, as well as reducing the soil's long-term ability to store water.

Plants have evolved some remarkable strategies that allow them to avoid or tolerate the pressures of grazing. Shrubs often have spines that deter grazers, and some grasses reduce their palatability by developing abrasive, tough, or bitter-tasting foliage. Plants may also escape notice by growing below grazing height, or rising high enough to escape heavy browsing.

Some plants have developed ways to tolerate what can appear to be destructive levels of grazing. Many grasses, for instance, have evolved rhizomatous root systems – networks of underground stems – that protect buds, store energy, and allow for rapid regrowth after grazing or trampling. These adaptations also help the plant to spread extensively, and protect it from being fully uprooted when grazed.

When grazing becomes excessive or prolonged, ecosystems can swiftly reach a point of collapse. The South American Pampas, for example, evolved without domestic grazers before Spanish colonization. Overgrazing in this landscape threatens the resilience of its plant communities, as well as a potential increase in invasive non-native species.

Conversely, when grazing stops altogether, grasslands can change dramatically. Without grazers, taller and woodier plants often take over, overshading and outcompeting species adapted to grazing. Unlike the Pampas, the vast Eurasian Steppe coevolved with grazing animals and is now threatened by changing land use and the absence of grazing animals. Restoration projects, such as the Altyn Dala Conservation Initiative in Kazakhstan focus on reintroducing keystone wild herbivores, such as saiga antelope or kulan (wild ass). The natural vegetation management of these animals, as well as their ability to disperse plant seeds and fertilize the soil through their droppings, is helping parts of the steppe become a diverse, healthy ecosystem once again.

Herd at work
The grasslands of the North American Great Plains struggled to cope with the grazing behaviour of dense herds of domestic cattle. The reintroduction of bison – which roam more widely – to replace cattle is proving effective at restoring a rich diversity of native plant species.

• GRASSLANDS AND SHRUBLANDS •

1/

Palmyra fruit is borne in clusters

1/ **Leaves in service**
Borassus flabellifer
The epitome of some savannas in South and Southeast Asia, the palmyra palm can grow up to 30m (100ft) tall. While it takes a long time to get going – it can spend its first 8 years developing a wide crown of leaves – it may go on to live for a further 100 years, with female plants producing an annual crop of 200–300 fruits when mature. Its fibrous, fan-shaped leaves are resistant to tearing and insect attack, and this durability has allowed them to be used like paper, as a writing material.

PLANTS OF
TROPICAL AND SUBTROPICAL GRASSLANDS, SAVANNA, AND SHRUBLANDS

Occurring in a wide band around the equator, tropical and subtropical grasslands, savannas, and shrublands typically receive relatively high amounts of precipitation but also experience strong seasonal dry periods. They are subject to regular disturbance because they usually have a continuous grassy groundlayer, which is both palatable to herbivores and flammable. Plants in this biome have evolved a variety of strategies to cope with drought, fire, and herbivory, including storing water, resprouting, and defensive structures such as spines.

TROPICAL AND SUBTROPICAL GRASSLANDS

2/

4/

3/

5/

2/ Shapeshifting pioneer
Myoporum sandwicense
Only found on the Hawaiian Islands, naio is sometimes termed false sandalwood, due to the scent of its wood. An extremely variable species depending on where it establishes itself, it may form a dwarf, prostrate shrub in lowland coastal shrubland, or grow into a large tree, up to 15 m (50 ft) tall, in upland habitats. This adaptable pioneer is often able to colonize new lava flows as they cool, helping to stabilize and enrich the soil for other plants, and laying the groundwork for new shrublands to develop.

3/ Resident introvert
Pilostyles blanchetii
An extraordinary plant without roots, stems, or even leaves, this species lives its whole life inside other plants, which it parasitizes for nutrients. Every year, its tiny brown or purple flowers and fruits emerge from within the stems of its hosts – mostly woody legumes – in the Campos Rupestres grasslands of Brazil.

4/ Lethal seeds
Sophora chrysophylla
Found only in the shrublands of Hawai'i, the bright yellow flowers of māmane distract from a deadly secret. Although the winged seedpods that follow the blooms are nutritious, they are also highly poisonous to all animals except the caterpillars of the cydia moth and the endangered palila bird, which has developed a tolerance to the seed's toxic alkaloids, as well as a taste for cydia larvae.

5/ Death benefit
Cynorkis flexuosa
On the dry grassy plateau of the highlands of central Madagascar, this orchid has found a way to survive in adverse conditions. When the landscape becomes dry, and the likelihood of fire increases, it becomes dormant and dies back. When environmental conditions improve, it uses energy stored in its woolly underground tubers to resprout.

6/ Fire protection
Pinus kesiya
The pine savannas of South and Southeast Asia are dominated by the khasi pine, which has a tall, straight trunk containing deep longitudinal fissures. It can reach heights of 35 m (115 ft), but it may be branchless for the first 20 m (65 ft) – an adaptation that protects the living crown from fires at ground level. Mature trees can also shed fire-damaged bark to allow for regrowth beneath.

Male cones are cylindroid and change from yelllow to light brown as they mature

6/

7 /

9 /

10 /

8 /

7/ Frequent fire-starter
Themeda triandra
Native to Africa, Asia, and Australia, red oat grass is a keystone species of tropical savannas. Its dense, tufted growth habit promotes regular, low-intensity fires that sustain the grassland structure and prevent woody plants from encroaching. Many grassland plants depend on this fire cycle to reproduce, relying on the heat to trigger seed release and germination.

8/ Two-tone trunk
Eucalyptus miniata
The iconic Darwin woollybutt inhabits the eucalyptus savanna woodlands that occur in a wide band around the north of Australia. Its common name derives from its striking bark, which is rough, fibrous, and grey to reddish brown at the base of the trunk – the "woolly butt" – and smooth, white to pale grey on the upper trunk and branches. The thick woolly bark protects the base of the tree from fire, providing an insulating layer that stops heat reaching living tissues within the tree trunk. Even if the bark chars, the inner layers remain much cooler.

9/ Flexible flora
Alysicarpus longifolius
In adapting to the variable environmental conditions that it experiences across the tropics, alyce clover has developed an ability to live either as an annual or as a perennial plant. When conditions are too dry, it will set seed and die off. However, when there is sufficient precipitation, the plant resprouts from a woody base. Perennial forms that develop deep woody crowns are even able to survive fire.

10/ Subterranean branches
Parinari capensis
The increase in global fires around four million years ago may have played a central role in the evolution of the dwarf mobola plum. Across tropical African grasslands and savannas, it has adapted to survive surface fires by growing an extensive network of underground stems and branches from which it can easily resprout.

11/ Twist and sprout
Heteropogon contortus
Black speargrass is a pivotal species of tropical savannas, where it is distributed from Africa to East Asia. Its seed features a bristlelike appendage called an awn, which can be up to 12 cm (5 in) long. The awn twists as a response to wetting and subsequent drying. This twisting action drills the end of the seed into the soil or into animal fur, which helps black speargrass to disperse – and plant – its seeds effectively.

11a / *Long, twisted awn forms a spear shape*

Awn straightens out as it takes in moisture

Awn contracts as it dries, resulting in a corkscrew motion

11b /

Wet and dry motion
Seeds of black speargrass are propelled into the ground and partially buried by a combination of the twisting movement of the drying awn, and the point at the seed's base, which acts like a drill bit.

Long green needles are arranged in fascicles (bundles) of two or three

12/ Fire blanket
Pinus yunnanensis

The Yunnan pine is native to four provinces in China, but found most widely in the province of Yunnan. Juvenile plants often go through a "grassy" stage, during which they develop a dense tuft of long, flexible needles, a structure that forms a shield to protect the tree's young growing tip from fire. Additionally, the plant's bark swells when it is exposed to fire, which enhances its fire-resistant properties and helps to create a thick layer that insulates the living tissues from heat.

13/ Small stabilizer
Vietnamosasa pusilla

In much of mainland Southeast Asia's dry dipterocarp savanna, this diminutive bamboo has found a home in the ground layer amongst the deciduous tree-grass vegetation. One of the smallest bamboos, often less than 1.5 m (5 ft) tall, it has extensive rhizomes that spread into the ground. The rhizomes help it to stabilize sandy and poor soils and resist drought, allowing it to grow in seasonally dry habitats where it is uncommon for other bamboos to survive.

14/ Essential support
Uapaca bojeri

The tapia tree is found only in Madagascar, where it is a key element of the Tapia ecosystem of the central highlands. In addition to shaping the local microclimate, and preventing soil erosion, this tree maintains biodiversity by sheltering unique plant, bird, and animal species, including lemurs. Most importantly, it is the only host plant for the Malagasy silkworm, a moth species whose cocoons provide the wild silk used in traditional textiles.

15/ Subterranean survivor
Andira humilis

Unusual in a group that is mostly made up of closely related rainforest trees, this species has found a novel way to survive frequent savanna fires. Native to Brazil, Bolivia, and Paraguay, it has moved most of its woody parts below ground, so that only leaves and short woody shoots protrude from the soil. Repeatedly checked by fire, these above-ground stems die and resprout over and over again, creating a low, shrubby habit. This life form, known as an underground tree, is a common adaptation when habitats undergo a regular disturbance, such as fire, frost, or heavy grazing.

16/ Rarely recorded
Paepalanthus gentlei

The genus *Paepalanthus*, known as pipeworts, includes many species that are microendemic in South America, which means they have a very small native range – in some cases, just a single mountaintop. This species is similarly restricted as it is only found in the grassy savannas of Belize. A small herbaceous plant that forms a tuft of rosetted leaves, it is so rare it has only been recorded a few times.

17/ Orchid hotspot
Cynorkis melinantha

This endangered orchid, which produces dense clusters of up to 20 striking yellow blooms, grows only in the open montane grasslands of central Madagascar. It thrives in a special ecosystem created by fire-adapted tapia trees, alongside many other orchids in the same genus. Around 90 per cent of *Cynorkis* species are endemic to this region, tiny areas of which are known as "nano hotspots" due to the high concentration of these co-existing orchids.

• TROPICAL AND SUBTROPICAL GRASSLANDS •

13 /

15 /

16 /

17 /

14 /

Round pendulums of buds hang on long stalks

Baobab tree

Adansonia digitata

Perhaps the most iconic species of the African savanna, the baobab tree (*Adansonia digitata*) is huge and majestic, reaching some 25 m (82 ft) in height, but also in girth. It is the longest-living angiosperm tree – some individuals have survived for more than 2,000 years – and has come to be known as the "Tree of Life". In addition to its incredible longevity, the tree has played an important part in sustaining the lives of others. There are more than 300 known uses – the leaves and fruit can be used in medicines, and the fruit contains 7–10 times more vitamin C than an orange. In fact, every part of the baobab is used as a source of food, medicine, or for spiritual welfare.

Baobab trees have a significant cultural history; for many, the tree is a symbol of longevity and wisdom. Rural communities revere the tree for its spiritual presence – for the Serer of Senegal the baobab is both a gathering place and a tomb for the knowledge bearers of the community. The importance of some individual trees has been formally recognized: one example is the Ombalantu baobab tree in Outapi, Namibia. A door leads to a space inside its hollow trunk that has been variously used as a chapel and a hiding place during Namibia's fight for independence, and it is now a national heritage site.

Flowers open at dusk to be pollinated by bats

"Wisdom is like a baobab tree, no one individual can embrace it."

Akan and Ewe proverb

Hand-sized leaves have 5–7 leaflets

Large, elongated fruits have a hard woody shell

Chapter 3
DESERTS

Rare rainfall

By definition deserts are dry with very little rainfall. However, many will experience short irregular rains due to fluctuations between El Niño and La Niña. Plants in coastal deserts rely on capturing moisture from fogs.

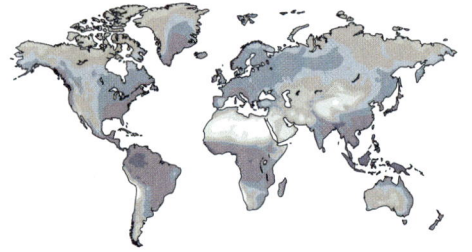

RAINFALL (MM)

- ○ 0–25
- 25–100
- 100–200
- 200–350
- 350–500
- 500–1,000
- 1,000–2,500
- 2,500–5,000

DESERTS

An extremely arid climate makes deserts one of the harshest environments for plants. Deserts also present other challenges such as extremes of temperature – typically hot daytimes or, outside the lowland tropics, freezing nights.

Winter frosts are a feature of the Great Basin Desert – more frequent that in the Mojave and Sonoran to the south

About 15 per cent of plants in the Caatinga are found nowhere else on Earth

The cold Humboldt current causes seasonal fogs in the coastal Lomas

KEY
- Coastal desert
- Hot, arid desert
- Semi-arid desert
- Cold desert
- — Average high °C
- — Average low °C

EXTREME CONDITIONS

Hot deserts are characterized by a formidable combination of drought and heat. Seasonal deserts can be hot and semi-arid with short rainy seasons – but away from the tropics, cold deserts typically experience mild summers and freezing cold winters. A rare type of desert is found in South America and southern Africa, where hyper-arid coastal deserts receive minimal amounts of moisture from ocean fog.

Highs and lows

Coastal deserts have fairly stable temperatures that are constant across the year, and through night and day. Hot, arid deserts and semi-arid deserts have hot days but cool nights. Cold deserts have significant seasonal temperature variations, and bigger fluctuations in day and night temperatures.

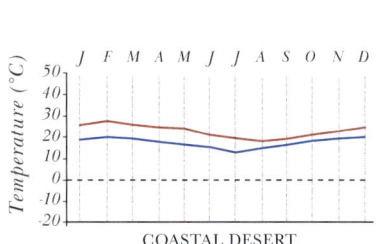

COASTAL DESERT

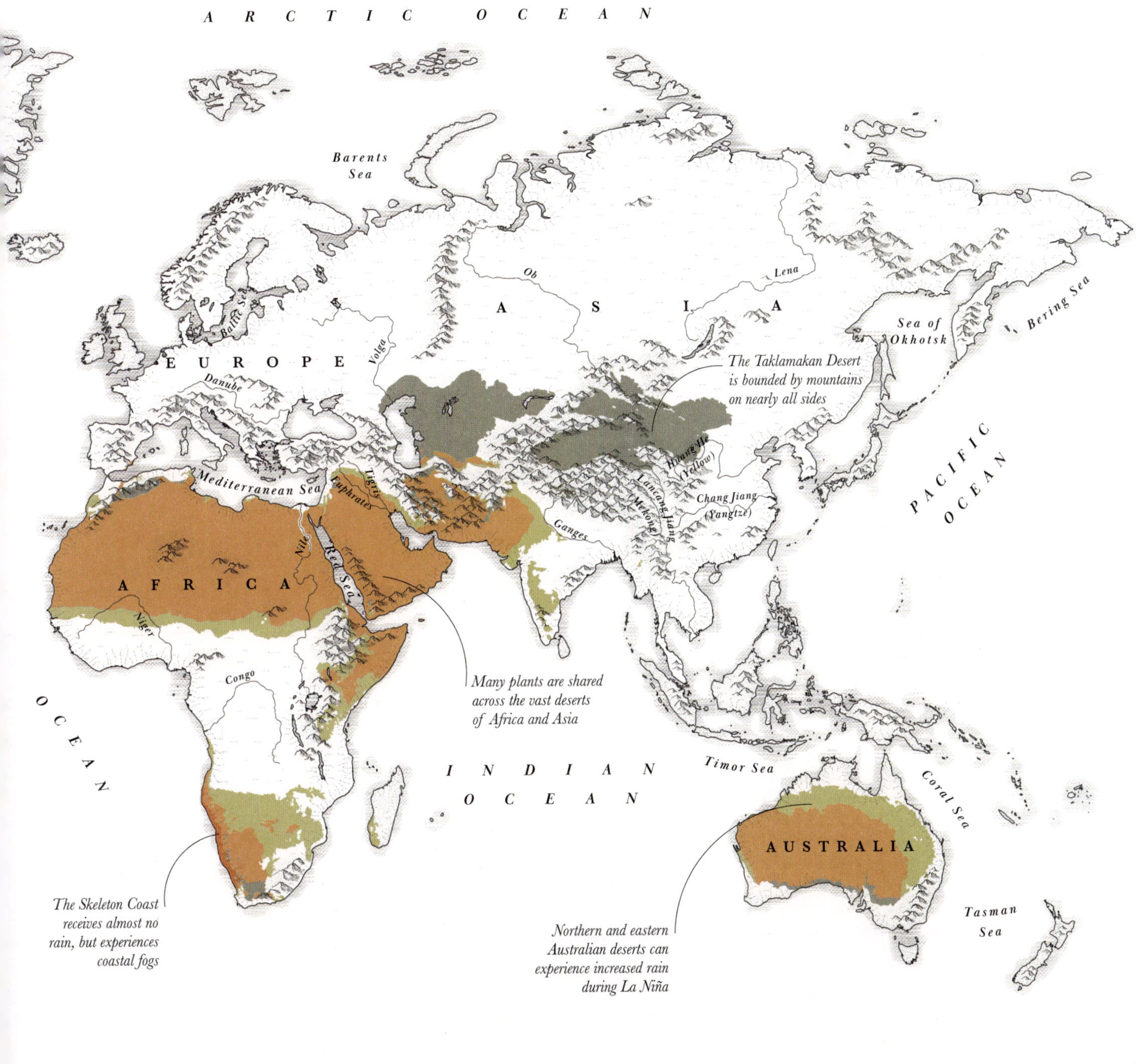

DESERTS

The Taklamakan Desert is bounded by mountains on nearly all sides

Many plants are shared across the vast deserts of Africa and Asia

The Skeleton Coast receives almost no rain, but experiences coastal fogs

Northern and eastern Australian deserts can experience increased rain during La Niña

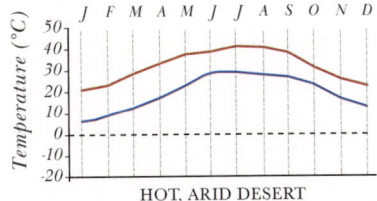
HOT, ARID DESERT

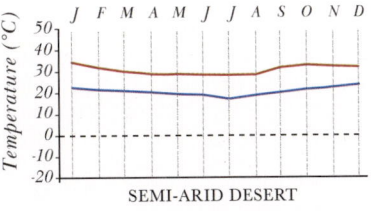
SEMI-ARID DESERT

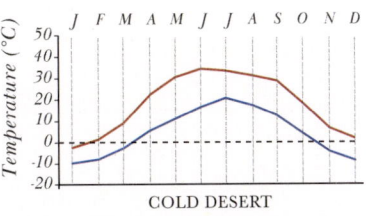
COLD DESERT

ADAPTATIONS
FORM AND WATER

Capturing, storing, and defending water is a fundamental part of survival in all deserts. Many specialist adaptations — including the overall growth forms of desert dwelling plants — revolve around capturing and retaining water because it is the main limiting resource. Succulent leaves — or losing the leaves entirely and replacing them with succulent, photosynthetic stems — are common strategies, and almost every green surface is covered with thick, waxy, water-retaining cuticles. Generally, sunlight is in abundance, so stems are often strongly erect to reduce overheating, and their slow growth rates allow them to accommodate periods when organic matter is scarce.

fig. 1 *Euphorbia polygona*; fig. 2 *Puya weberiana*; fig. 3 *Cereus peruvianus*; fig. 4 *Agave deserti*; fig. 5 *Mesembryanthemum crystallinum*; fig. 6 *Hoodia gordonii*; fig. 7 *Opuntia polyacantha*; fig. 8 *Ferocactus glaucescens*; fig. 9 *Cereus lamprospermus* subsp. *colosseus*; fig. 10 *Euphorbia ingens*; fig. 11 *Fouquieria splendens*; fig. 12 *Agave victoria-reginae*; fig. 13 *Phoenix dactylifera*

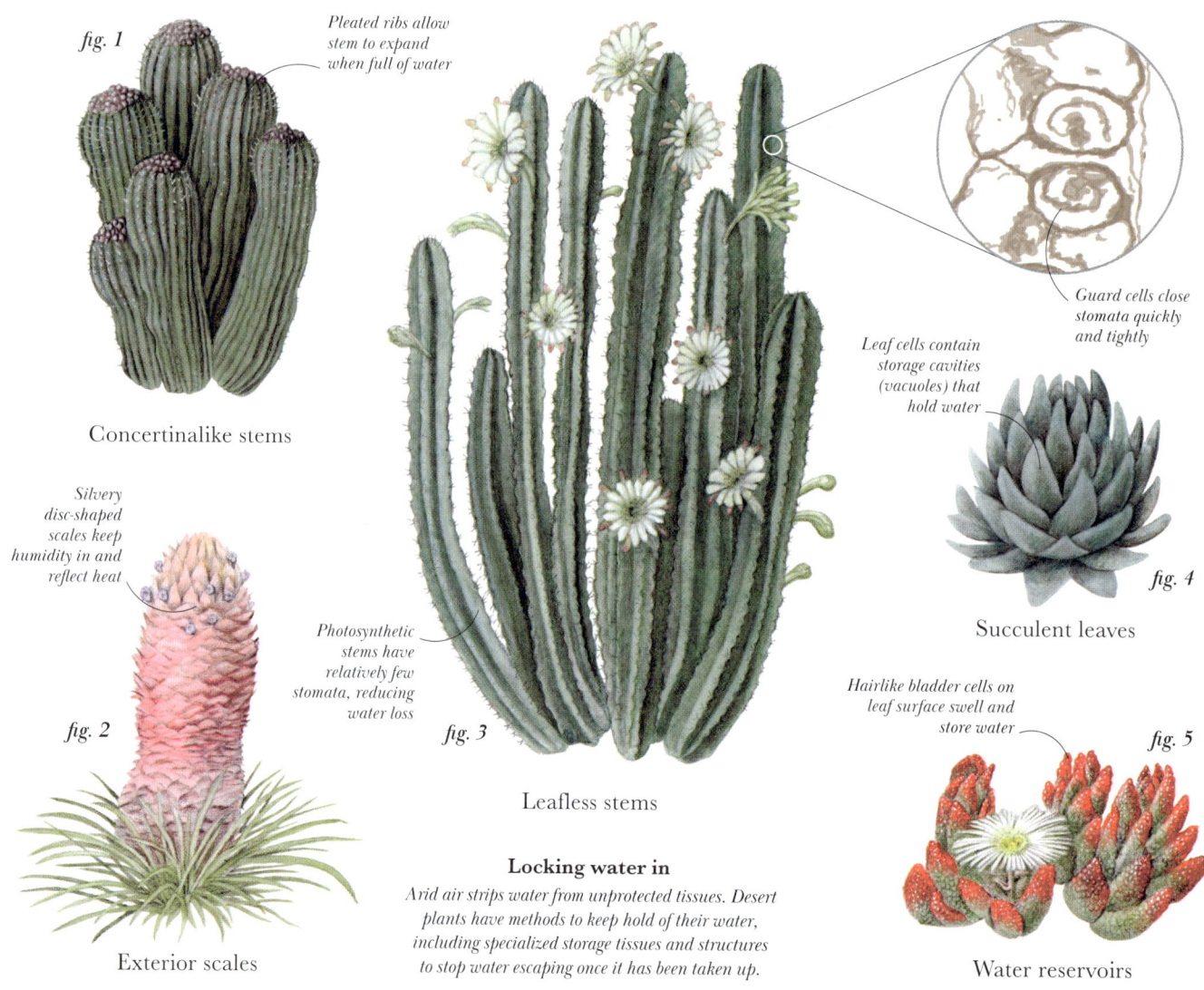

fig. 1 — Pleated ribs allow stem to expand when full of water
Concertinalike stems

fig. 2 — Silvery disc-shaped scales keep humidity in and reflect heat
Exterior scales

fig. 3 — Photosynthetic stems have relatively few stomata, reducing water loss
Leafless stems

Guard cells close stomata quickly and tightly

fig. 4 — Leaf cells contain storage cavities (vacuoles) that hold water
Succulent leaves

fig. 5 — Hairlike bladder cells on leaf surface swell and store water
Water reservoirs

Locking water in

Arid air strips water from unprotected tissues. Desert plants have methods to keep hold of their water, including specialized storage tissues and structures to stop water escaping once it has been taken up.

• DESERTS •

Clumps of stems form colonies with relatively moist centres

fig. 6
Colonies

fig. 7

Pads that fall to the ground can root and create new clones

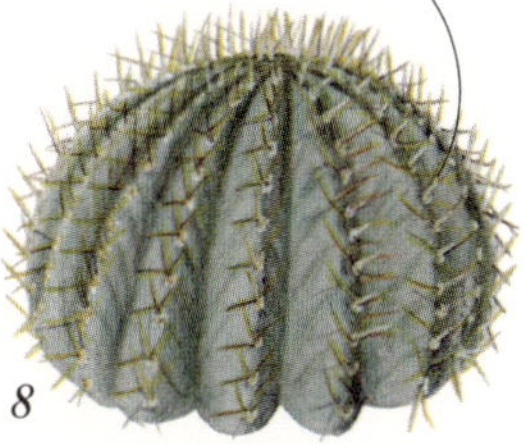

Swollen, barrel-like stems store large quantities of water

Padlike forms

fig. 8
Barrel

Dew condenses on the hard ridges of leaves

Water runs down into a space-efficient spiral of leaves

fig. 12
Spiral funnels

Stems are raised high to avoid browsers

Upright stems may reach heights of 15 m (49 ft)

Succulent leaves sit on slender, woody stems

fig. 10
Treelike

fig. 9
Columnar

fig. 11
Spiny canes

Growth habits

Desert plants have developed an array of different forms and growth habits – barrels, columns, and colonies – that are supremely adapted to collecting and keeping water.

Leaves can absorb condensation from the air due to shade

Upper roots spread horizontally to catch surface water or dew

fig. 13

Deep roots can reach water up to 5 m (16 ft) below the surface

Deep and wide root system

Capturing water

Water in deserts can come from rare rainfall or from condensing dew. However, plants also extend their roots long distances below ground level to search out groundwater.

Pollination

Desert plants deploy a variety of pollination strategies. They range from the loose and general, designed to attract a range of pollinators, to precise methods, such as relying on a single specialist insect or animal species to fertilize their flowers – a risky tactic should the pollinators disappear. Some plants have a backup option and allow their flowers to pollinate themselves.

Large floral throat suits many types of visitors, including large bat species

fig. 1

Oversized flowers

Self-incompatible blooms cannot fertilize themselves; most of these flowers receive pollen via animals that travel from one plant to another

Showy, bright-white tepals attract birds, bees, and bats to nectar day and night

fig. 2

Bright, open flowers

Yellow, narrow-necked flowers are adapted to accommodate both bee and sunbird pollinators

fig. 3

Dual pollinators

Many stamens produce copious amounts of pollen to ensure pollinators are completely covered

fig. 4

Copious pollen

Flowers produce scent compounds to attract a specific species of yucca moth

fig. 5

Strong relationships

Female stigma

Pollen is transferred from stamens to stigma within the flower if no pollinators visit

Male stamen

fig. 6

Self-pollination

• DESERTS •

ADAPTATIONS
POLLINATION AND DEFENCE

Individual desert plants are often scattered far and wide, so – unlike in grassland – wind pollination is seldom a viable strategy. Instead, animal pollination is the norm. However, pollinators are relatively scarce in deserts, so flowers are often strikingly attractive and colourful to encourage visits.

Interactions between desert plants and animals are not always friendly. The valuable trove of water and nutrients that the plants hold makes them targets for herbivores who can satisfy their hunger and slake their thirst in one easy meal – or could if the plants did not have such sophisticated defences.

fig. 1 Echinopsis candicans; *fig. 2* Carnegiea gigantea; *fig. 3* Aloe kraussii; *fig. 4* Mesembryanthemum digitatum; *fig. 5* Yucca brevifolia; *fig. 6* Cistanthe longiscapa; *fig. 7* Cochemiea poselgeri; *fig. 8* Aloe aculeata; *fig. 9* Opuntia basilaris; *fig. 10* Euphorbia ammak

Spiny defences
Desert plants have developed numerous spiky ways to defend themselves, and protect their precious water and resources from animals. Plants without spines usually have another trick in the form of virulent toxins.

Hooked spines snag animal fur to disperse cactus segments

Paired spines evolved from stipules rather than from buds

Buds are modified into spine clusters called areoles, with a large spine in the centre

Spines grow on the margins and surface of succulent leaves

Spines are small and relatively sparse as the plant is protected by caustic poison

Tiny barbs at the base of areoles lodge in animal flesh

fig. 7 Hooked spines

fig. 8 Edge spines

fig. 9 Barbed spines

fig. 10 Paired spines

PLANTS OF
HOT DRY DESERTS

Shaped by a brutal combination of arid climates and high daytime temperatures, hot, dry deserts occur across every major continent, generally following the tropics of Cancer and Capricorn. Skies remain cloudless for months, creating dramatic swings between hot days and cool nights. Despite these extremes, many plants survive through remarkable adaptations: storing water in swollen stems, reducing leaves to spines, or waiting in suspended animation as seeds until rare rains arrive. Together they show how life endures in some of the harshest conditions.

1 /

2 /

4 /

Thorns help to protect the myrrh tree from grazing animals

3 /

1 / Resisting heat
Larrea tridentata
In the Mojave Desert, the creosote bush protects itself from fierce heat by having small, resin-coated leaves sealed with a waxy varnish to curb water loss. Its branches are strategically angled to reduce exposure to the full intensity of the afternoon sun. To preserve their energy resources in the difficult climate, creosote bushes are slow-growing and reproduce clonally with a ring of genetically identical plants forming around the parent. The oldest clonal colony is thought to be some 11,700 years old.

2 / Alien flowerheads
Dorstenia foetida
A relative of the fig, grendelion survives arid conditions in Northeast Africa and the Arabian Peninsula by having a short, swollen stem base that can store water like a bottle. Its unusual fleshy flowerheads are star- or disc-shaped, resembling an alien spaceship, and contain tiny male and female flowers, followed by fruits. When ripe, these fruits explode, forcefully ejecting the seeds.

3 / Aromatic defence
Commiphora myrrha
Native to the arid hills of the Horn of Africa and the Arabian Peninsula, the myrrh tree survives on scant rainfall by storing water in its swollen trunk and shedding leaves during prolonged drought. Its bark exudes aromatic resin that seals wounds and deters herbivores. This fragrant substance has also been prized for use as incense and in medicine.

4 / Aged survivor
Pinus longaeva
The Great Basin bristlecone pine has endured millennia on the harsh, rocky slopes of the western US. It survives by growing incredibly slowly, forming dense, resinous wood that resists decay, insects, and disease. Some trees are confirmed to be over 4,800 years old, which means that they were already growing when the Great Pyramids were built.

• HOT DRY DESERTS •

5/ Giant of the Sonora
Carnegiea gigantea
The saguaro cactus of the Sonoran Desert in North America is a master of desert survival, storing vast quantities of water in an expandable, pleated stem, which can measure 12 m (40 ft) tall. It provides food and shelter to dozens of species of birds, mammals, reptiles, and insects. Opening at night, its nectar-rich flowers are specially adapted to attract bat pollinators, but they also sustain bees and hummingbirds at dawn and dusk.

Each blossom lasts less than 24 hours before wilting

6 /

7 /

8a /

9 /

10 /

• HOT DRY DESERTS •

6/ Living off others
Tapinanthus oleifolius
The namnambush is one of many hemiparasites found in arid regions. These plants can photosynthesize but top up their resources by parasitizing others. Like the desert hyacinth, the namnambush uses a special rootlike organ (haustoria) to attach itself to a host and extract the nutrients it needs. Birds pollinate the flowers and later deposit the plant's seed on the branches and stems of new hosts, where they quickly germinate and form new namnambush plants.

7/ Staying small
Aluta maisonneuvei
Across the arid deserts of western Australia, the desert heath myrtle copes with intense heat and drought by having tiny leaves, only a few millimetres long, to reduce water loss. Even its flowers have evolved to be so small that they only contain half the number of stamens that most members of this plant family would normally have – a miniaturization that saves water and energy in this harsh landscape.

8/ Swift revival
Selaginella lepidophylla
The resurrection plant from the Chihuahuan Desert in Mexico and US can survive extreme droughts despite completely desiccating and appearing dead. Its cells contain protective sugars and proteins that prevent damage during the drying process. When air humidity increases before rainfall, or the plant comes into contact with water, it springs back to life, the stems turning green as they rehydrate.

9/ Forest relic
Dracaena cinnabari
The dragon blood tree of Socotra, a remote island in the Arabian Sea, survives thanks to its distinctive umbrella-shaped canopy, which captures moisture from fog and shades its roots. It is thought to be a remnant of the subtropical forests that once covered much of North Africa. As the region became desert, these forests vanished, leaving the dragon blood tree's closest relative – the similar-looking *Dracaena draco* – stranded more than 7,500 km (5,000 miles) away on the Canary Islands.

10/ Hidden beauty
Cistanche tubulosa
Found in the most barren parts of Europe, Asia, and Africa, the desert hyacinth survives through parasitism. Lacking chlorophyll, it cannot photosynthesize, so uses a special rootlike organ to penetrate the tissue of a host plant to extract water and nutrients. This hidden lifestyle allows the desert hyacinth to thrive where few green plants can, revealing itself only as a golden flower spike above the sand.

11/ Desert dyes
Lawsonia inermis
Thriving in the hot, dry soils of North Africa, the Arabian Peninsula, and Southwest Asia, the henna tree has deep roots and small, tough leaves. The leaves help it to withstand intense heat and drought. The leaves produce protective pigments that may help shield the plant from sunlight and pests. When dried and powdered, these pigments yield the reddish dye long used to colour skin, hair, and fabrics – a tradition that dates back to Ancient Egypt.

Small, glossy leaves reflect sunlight to limit heat stress

11 /

Fronds uncurl and turn green within hours of contact with water

Tightly curled, dry fronds protect inner tissues

8b /

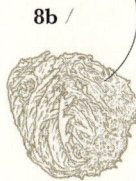
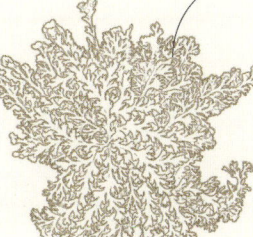

Before   After

Unfolding after rain
Rehydration restores cell pressure in the plant, and softens its rigid stems. The dry rosette loosens, unfolding into a green, active plant capable of photosynthesis once more.

12/ Sacred resins
Boswellia sacra
In the rocky deserts of Northeast Africa and Arabia, the frankincense tree survives where little else can. Its thin, peeling bark and drought-tolerant stems help it conserve water, while its shallow roots capture scarce rainfall. Like many members of the Burseraceae family, frankincense trees have aromatic compounds that seal wounds to their trunks or branches, and also protect against infection. These resins are tapped by cutting the bark and have been traded for use as perfumes, incense, and medicine for thousands of years.

13/ Blooming bottles
Adenium obesum
The swollen trunks and protruding roots of the desert rose are the classic example of a "bottle tree", a succulent form of tree found in deserts in Africa and the Arabian peninsula. The stored water in their trunks sustains them through long, rainless seasons in extreme desert heat. Even a little rain induces these striking trees to produce a flush of pink trumpet-shaped flowers.

14/ Seeds in waiting
Geraea canescens
One tactic to endure years without rain in the extreme heat of Death Valley, US, is to survive as hard seeds. In this inhospitable environment, the seeds of ephemeral flowers such as desert gold lie dormant until rare rains arrive. Within weeks of their arrival, the desert gold seeds burst into life, carpeting otherwise barren plains with golden blooms before vanishing again, leaving only seeds behind.

15/ Ancient isolation
Macrozamia macdonnellii
The MacDonnell Ranges cycad grows in the vast desert that covers the centre of Australia's Northern Territory. It survives in a few sheltered pockets around rocky outcrops, where shade and higher humidity protect it from the extreme desert heat. It is thought that Macrozamia cycads were once widespread across a wetter Australia, but as the continent dried, the MacDonnell Ranges cycad became isolate from its nearest relatives by more than 1,300 km (800 miles).

16/ Half a flower
Scaevola spinescens
Widespread across the deserts of Australia, the spiny fan-flower has an unusual flower form with all petals on one side, as if the bloom has been cut in half. This special adaptation ensures that visiting pollinators touch both the anthers – which deposit pollen – and the stigma – which receives pollen – improving the chances of pollination. It has long been valued by the Noongar people of southwest Australia for its medicinal uses.

17/ Desert oddity
Fouquieria columnaris
One of the strangest examples of stem succulence can be found in the hyper-arid deserts of Baja California, which receive less than 140 mm (5 in) of rain a year. The boojum tree has evolved a single, tapering stem that is supported by the water pressure within its tissues rather than by woody xylem. This unique construction allows the plant to store water efficiently and survive long, rainless years.

Leaves are shed quickly during droughts to conserve moisture

12 /

• HOT DRY DESERTS •

13 /
14 /
15 /
16 /
17 /

CACTI

The cactus family (Cactaceae) diversified from its South American origins with the formation of arid habitats as the Andes rose. New species developed as plants spread and as pollinators changed.

Cacti evolved in a patchwork of dry habitats. From there they spread across the Americas, then travelled with humans to other continents. Mesic cacti, which grow with a moderate amount of moisture, were able to spread directly through wet forests, while arid cacti have been more successful diversifying and dominating desert ecosystems. Most are vulnerable to cold, but a few species in cold deserts tolerate frost.

HISTORIC MIGRATION
- Arid cacti migration →
- Mesic cacti migration →
- Human intervention →

REPRESENTATIVE LIVING SPECIES
- *Carnegiea gigantea*
- *Cleistocactus baumannii*
- *Epiphyllum oxypetalum*
- *Opuntia fragilis*
- *Opuntia galapageia*
- *Rhipsalis baccifera*

Cleistocactus baumannii
The firecracker cactus is a rare example of a pollinator specialist cactus. Its long narrow, red, tubular flowers are only visited by hummingbirds.

Thick flower tube and protective bracts stop nectar robbers

Spineless stems, typical of many epiphytic cacti

① The cactus family evolves 30–35 million years ago with the formation of the Andes in the area of modern-day Peru. This region still has the highest cactus diversity.

② Arid-specialist cacti move into deserts across South America, forming a second centre of diversity in northeast Brazil.

③ Arid cacti move over the Panama isthmus land bridge into the deserts of Mexico and southwestern North America.

④ Mesic cacti spread north through the wet forests of Central America and the Caribbean.

⑤ Rhipsalis originates in Paraguay and northern Argentina, then spreads to the Atlantic coast of Brazil, from where it moves to the Caribbean and North America.

⑥ *Rhipsalis baccifera* is thought to have been carried across the Atlantic by birds to tropical Africa, Madagascar, and Sri Lanka.

• DESERTS •

Rhipsalis baccifera
Mistletoe cactus is the only cactus to occur naturally outside the Americas. Its small white fruit are eaten by birds, who also act as seed dispersers.

Epiphyllum oxypetalum
An epiphytic cactus, queen of the night has large, nectar-rich flowers that open for a single night to attract the bats and moths needed for pollination from a long distance.

Brilliant white blooms are highly fragrant

7 Humans spread cacti across the world. Opuntia is probably the most widely grown and a common sight in arid climates, although it is considered invasive in places.

Flowers that are pollinated turn into dry fruits, full of seeds

Opuntia fragilis
A low-growing cactus, brittle prickly pear reaches as far as 56°N. It survives the harsh North American winters by hiding under snow cover and reducing water content to avoid frost damage.

Holes excavated in stems by woodpeckers are used as nests by owls and wrens

Long trunk raises pads and fruit above browsing height

Opuntia galapageia
This flexible and variable prickly pear from the Galápagos has adapted to a range of conditions. Normally a ground-hugging shrub, when it coexists with giant tortoises it grows as a tree.

Carnegiea gigantea
Saguaro flowers unfurl at night to attract bat pollinators, but birds and insects also visit before dusk and after dawn while the flowers are still open. Many vertebrates eat the fruit.

Prickly pear

Opuntia ficus-indica

Prickly pears (*Opuntia* sp.) were domesticated and grown across much of Central America long before Europeans arrived, and have since spread across the world. They are found in arid regions where they are used for food and to mark property boundaries. The importance of the prickly pear to the people of Central America is reflected in the variety of common names: most Spanish speakers will know it as *nopal*, which comes from the Nahuatl word *nohpalli*; there is also a range of names across Mayan languages including *nach'te* in K'iché Maya, and *pa'kam* in Mopan Maya.

Perhaps surprisingly, both the fruit and the spiny pads are eaten. The pads can be cooked in a variety of dishes including salads, stews, and tacos. The fruits, known as *tuna* in Spanish, are popular too, with crisp, sweet flesh full of tiny seeds.

The prickly pear is also a key part of traditional medicine among Indigenous communities in Central America where it is used to treat a range of maladies – from headaches and fever to diabetes and arthritis. Such medicinal uses have been adopted by other groups in the region, including the Creole people of Belize who call it *scoggineal*. Widely known and valued for its myriad uses, the prickly pear has become a symbol of food security and survival for many communities.

Founding Mexico City
The Mexica people founded Tenochtitlan – now Mexico City – after witnessing an eagle eating a snake while perched on a prickly pear. The event fulfilled a prophecy by the god Huitzilopochtli and is now an emblem for the city.

Flowers fade and mature into fruit

> *"It is round, with a top like a spindle whorl, the top filled out, round, slender-based. It is prickly, full of thorns."*

Bernardino de Sahagún, *Florentine Codex*, Book 11, 1540–1585

DESERTS

Ripe fruits have green or pinky-red flesh

Succulent pads – or cladodes – store water, and also perform photosynthesis

Seed capsules have 2–3 papery wings

Female flowers catch pollen from male plants on the wind

1/ Global traveller
Dodonaea viscosa
The akeake is a hardy shrub with sticky, resinous leaves that produce chemicals, which inhibits the growth of nearby plants. Its papery, winged seed capsules hitch a ride on the wind and can travel vast distances. This has helped the unassuming akeake to become one of the most widespread species in the world – it occured in arid regions across the tropics and subtropics, as well as in temperate habitats. Different populations can look remarkably different, as it shows huge variation in form across its range.

PLANTS OF

SEMI-ARID DESERTS

Found predominantly across the Tropics of Cancer and Capricorn, semi-arid deserts and often form a boundary biome between hot deserts and wetter biomes. These regions experience extremely seasonal rainfall, usually one to three months of rain, followed by a prolonged dry period and high daytime temperatures for the rest of the year. Like hot, arid deserts they rarely experience frosts, but semi-arid skies have cloud cover during the brief rainy seasons, that allows the plants in these biomes to burst into life with sudden flushes of leaves and flowers.

• SEMI-ARID DESERTS •

2 / Moroccan goat trees
Sideroxylon spinosum
The argan tree of southwestern Morocco is adapted to dry, rocky soils, where its deep taproots draw up water from far below the surface. The tree's oil-rich fruits are eaten by goats, which disperse the seeds, enabling new trees to take root across the arid landscape. The tree's extensive root system stabilizes the soil and prevents wind erosion, helping to combat the desertification of the region.

3 / Primate pollinators
Adansonia grandidieri
An iconic Madagascan baobab, the reniala has large white flowers that open at night for pollination by hawkmoths and bats. But in a twist unique to Madagascar, the flowers are also pollinated by lemurs. As they climb through the canopy, licking nectar out of each cuplike flower, the lemur faces become covered in pollen, which is passed to the next flower. Once pollinated, the tree produces large, woody fruits.

4 / Feeding the forest
Diospyros lycioides
Like other members of the ebony genus *Diospyros*, the bluebush from Africa's dry forests produces exceptionally dense, dark wood and tough leaves. Denser wood reduces water loss and protects the plant from the harsh sun. The bluebush is a key part of the ecosystem; its fruits are eaten by a range of birds and animals, as well as humans, the Mooi river opal butterfly feeds on its leaves.

5 /

6 /

7 /

8 /

9 /

SEMI-ARID DESERTS

5/ Unfussy benefactor
Grevillea wickhamii
The holly-leaf grevillea is a shrub from the arid regions of northern Australia. Much of its success is down to its long clusters of yellow and red flowers. Each flower consists of a long tube containing abundant nectar for visitors with long tongues. These flowers are not fussy – they can feed and be pollinated by a range of creatures including honeyeaters and butterflies. It blooms when little else does, which makes the shrub an important food source in the dry season.

6/ Spiny mouthful
Ziziphus nummularia
Found in the Thar Desert in northwestern India and eastern Pakistan, the ber survives by proving simply too difficult to eat. It holds its leaves within a tangle of woody zigzagging branches. The branches are further protected by pairs of long woody spines at each node, that keep even the most determined browser away from its leaves. These defences allow the shrub to retain its foliage through long dry seasons, conserving moisture and surviving in exposed, overgrazed desert scrub.

7/ Fire-resistant bottle
Brachychiton rupestris
The Queensland bottle tree, found only in the open grasslands and dry forests of central and southeast Queensland, is an independent evolution of the succulent bottle tree habit in arid habitats. This tree is more unusual than its relatives because it tolerates fire – rare for a succulent plant – and is quick to resprout, making this bottle trees an important habitat species for the region's endangered birds.

8/ Succulent oddity
Didierea madagascariensis
The sono or Madagascan octopus tree is a curious-looking succulent tree from a small family found only in southwest Madagascar, where it is adapted to the arid spiny forest. Unlike many succulents, these trees have small leaves, which are hidden among the long trunk spines. The spines deter grazing animals and help to shade the trunk, reducing the amount of water lost in the intense dry heat.

9/ Tumbleseed
Boophone disticha
After rains, the century plant produces a dense, round head of sweetly scented pink or red flowers. The flower stalks lengthen as the seeds mature, transforming the flower head into a light, spherical structure. Eventually the whole inflorescence breaks off from its stalk to act as a tumbleweed, shedding mature seeds as it rolls across the landscape in the dry season breezes. This clever dispersal strategy allows the plant to spread its seeds widely across its home range in the arid regions of southern Africa.

10/ Sahelian umbrella
Vachellia tortilis
The distinctive silhouette of the umbrella thorn acacia is a familiar sight across the Sahel, a dry belt of savanna stretching across Africa south of the Sahara Desert. Its wide, flat-topped canopy shades its own roots and helps to reduce water loss from the soil below. For numerous birds and animals, the thorny branches provide welcome roosting and foraging sites in the harsh landscape. Birds also build their nests in the canopy and large herbivores feed on its nutritious bean pods.

10 /

Pairs of straight spines protect each leaf stalk

11/ Particular parasite
Krameria ixine

Extreme specialization rarely pays off in a desert. In dry forests and deserts across the Americas, the desert ratany survives by parasitizing the roots of nearby plants. This low-growing shrub can tap into dozens of different species, drawing water and nutrients from its hosts. Its small pink flowers, however, rely on a single group of pollinators – centris bees. These solitary bees collect oil to feed their young from blister glands on the petals as their reward.

12/ Separated by sea
Pachypodium geayi

This species from southwestern Madagascar represents one of several separate evolutions of bottle-shaped succulents in the dogbane family. The pale, columnar trunks that earned them their name store water to endure long dry seasons. Although *Pachypodium* species grow across Madagascar and southern Africa, their relationship remains uncertain, which suggests that their ancestors may have crossed the ocean long after the two lands separated.

13/ Fruit for bears
Colicodendron scabridum

The sapote tree grows in the dry forests of the valleys and coastal regions of northern Peru and southern Ecuador. Its large, hard fruits are consumed by various birds and mammals, which disperse the seeds to new locations. Andean spectacled bears that are normally found in humid cloud forests move to the valleys when the fruit is in season.

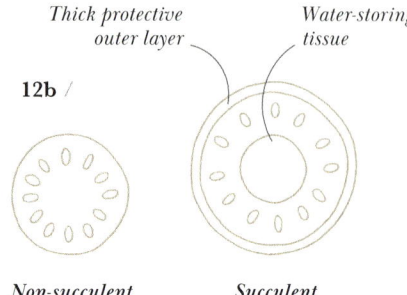

Thick protective outer layer *Water-storing tissue*

12b /

Non-succulent   *Succulent*

Storing water
The succulent stem of Pachypodium geayi contains thick, water-storing tissue to survive long dry periods. A tough outer layer reduces water loss.

14/ False leaves
Acacia holosericea

Like many Australian acacia species, the shrubby silver wattle of the northern semi-desert areas in Australia looks like a normal tree, but it does not have true leaves. Each "leaf" is a phyllode – a flattened and widened leaf stalk that carries out photosynthesis in place of the lost leaf. These phyllodes help the plant to survive heat and drought, making the modern plant's foliage far more resilient than its ancestor's.

15/ Lookalike cactus
Euphorbia caducifolia

Life in semi-arid deserts demands clever adaptations, such as storing water in thick stems and deterring grazers with spines. This is why across Africa and Asia, many species of *Euphorbia* closely resemble the cacti of the Americas. The leafless milk hedge (*E. caducifolia*) from semi-arid deserts in India looks like a cactus but can be easily distinguished: it produces milky white latex, which is never present in cacti; has two spines per node instead of many spines in a cactus; and bears small simplified flowers in structures called cyathia.

16/ Breakfast buffet
Banksia prionotes

The acorn banksia of southwestern Australia blossoms at dawn when honeyeaters – its primary pollinator – are most active as the birds prefer freshly opened flowers due to their higher nectar content. The flowers employ a clever two-stage system: first, they release their pollen, and only later do their female parts become receptive. This ensures that the first birds to visit carry pollen away, while later visitors deliver pollen from other flowers, which improves the likelihood of cross-pollination.

11 /

Spiny fruit capsules

Flowers have specialized glands that produce nutrient-rich oil

• SEMI-ARID DESERTS •

12a /

13 /

14 /

15 /

16 /

DESERT GREENING

One of the main predictions of global warming was that many dry places would get even drier and that deserts across the world would grow larger. However, it is clear from satellite imagery since the 1980s that some deserts and other arid areas are actually getting greener. Not only is vegetation expanding into deserts, but the plants that are moving there are bigger and leafier. These changes have been caused in part by higher levels of carbon dioxide in the atmosphere, which increases the rate of photosynthesis in plants.

During the day, plants allow carbon dioxide to enter their tissues through small pores called stomata to perform photosynthesis. The increasing supply of carbon dioxide, however, means that plants do not need to open their stomata quite as wide to take in the carbon they require for photosynthesis. With their stomata partially closed, plants are also losing less water through evaporation. Essentially, plants are using water more efficiently. This is stimulating the growth of vegetation at a rate that outweighs the negative effects of increased aridity caused by global warming. The net result is that some deserts are becoming greener.

While this may appear to be welcome news, in some areas, desert greening poses significant problems: animals that are specially adapted to arid conditions may come under threat, and the additional vegetation in arid parts of southeast Australia has increased the risk of brushfires. On the other hand, in areas such as Niger, trees have regenerated, supplying agricultural crops with nutrients and providing shade for animals. Recent climate modelling suggests that the rate of desert greening may not slow down before the middle of the 21st century – but whether or not this is a "net positive" is down to the sands of time.

Shifting sands
Trees planted across Mali were intended to slow desertification as overgrazing caused the Sahara Desert to expand southwards. Instead, as plants benefit from increased carbon dioxide, vegetation may also be spreading north in some areas, contributing to localized desert greening.

• DESERTS •

PLANTS OF
COASTAL DESERTS

Found in primarily two locations, coastal deserts are caused by cold upwelling currents offshore: the Benguela current off Namibia, and the Humboldt current off Peru and Chile. The deserts are among the driest places on Earth, and even include areas that have no recorded rainfall. However, these seemingly barren landscapes often have cloudy skies, and for several months of the year, cool fog drifts inland from the ocean, where it provides a welcome source of moisture. The native plants have evolved remarkable adaptations to harvest water from the air, and briefly burst into life.

Clusters of flowers offer rich nectar to sunbirds and other pollinators

1/ Reflecting heat
Aloidendron dichotomum
A striking succulent tree, the quiver tree is found along the rocky cliffs that track the Skeleton Coast in the Namib Desert. A notable adaptation is a light, powdery bloom and pale colouration, which reflects the intense desert sun, and stops the tree from overheating in the extreme heat. Its fleshy, spear-shaped, blue-green leaves are borne in rosettes at the end of the tree's branches.

2/ Bell flowers of the fog
Nolana paradoxa
Bell flowers (*Nolana* spp.) are a typical sight across Peruvian and Chilean Lomas – the isolated, fog-watered oases in the coastal desert. These succulents burst into bloom to capitalize on the precious moisture provided by the seasonal fog. All Lomas patches have at least one species of bell flower; some have as many as ten. Some species grow as annuals producing a new plant each year, others are perennials that survive underground between fog events.

Fleshy leaves store water collected from coastal fog

2 /

3a /

4 /

5 /

6 /

3/ Ancient enigma
Welwitschia mirabilis
One of the most bizarre and loneliest plants on the planet, the welwitschia is the only species in its genus, and has no close living relatives. Found only in the coastal fog belt of the Namib Desert, these plants consist of two giant leaves that grow continuously from the base. The leaves shred and split, which makes them look more numerous. Only the small cones give away their true gymnosperm identity. These plants grow so slowly that the largest individuals are thought to be thousands of years old.

4/ Spine-protected melon
Acanthosicyos horridus
In the Namib Desert the !nara melon, as it is known in a local Khoisan language, uses a deep taproot, reaching depths of up to 50 m (160 ft) to access groundwater. It also saves water by photosynthesizing with its green stems and spines instead of leaves. The !nara melon is considered a keystone species; its twisted, spiny stems stabilize sand, creating microhabitats for other plants, insects, and small mammals. Its heavy, spine-encased melons are eaten by jackals and other large mammals.

5/ Channeling water
Zygophyllum stapffii
The strange-looking dollar bush pokes its pairs of succulent leaves above the sands of Namibia's coastal desert. Moisture from the foggy atmosphere condenses on and runs down its leaves, dripping onto the sand where it can be absorbed by the plant's shallow roots. This ingenious self-watering system allows the dollar bush to thrive in an environment that receives almost no rain.

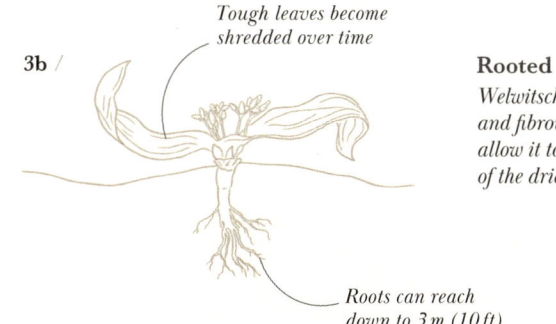

3b /
Tough leaves become shredded over time
Roots can reach down to 3 m (10 ft) below the surface

Rooted in extremes
Welwitschia's deep taproots and fibrous tough leaves allow it to survive in one of the driest places on Earth.

6/ Orchids of the fog deserts
Chloraea pavonii
The large, striking yellow flowers of the Lima orchid, which bloom during the seasonal mists from August to September, are commonly found in the fog-fed Lomas ecosystems near the Peruvian capital. The flowers have a labellum (landing platform) that is covered in toothlike projections, which are thought to act as sensory signposts, guiding pollinators towards the nectaries at the base of the flower.

7/ Living in two worlds
Begonia octopetala
A charismatic member of the megadiverse *Begonia* genus, *B. octopetala* has a surprising distribution. It can be found in coastal Ecuador and in the Lomas of Peru where its deep underground tuber waits for the annual fog seasons from August to September. The same species can also be found in cloud shrouded shrublands high up in the Andes where they flower in the wet season, from January to April.

7 /
Masses of flowers, each often with eight tepals, appear rapidly after seasonal moisture arrives

8/ Festive blooms
Ismene amancaes

The threatened amancae is a hardy bulb that survives long, dry months underground, before its bright yellow blooms emerge with precision for the fog season. These spectacular flowers were once so abundant that they were collected for Lima's Amancaes Festival in June. Due to over-collecting, the plant has vanished from its native hills around the city, and the festival that it inspired is now only a memory.

9/ Dune daisy
Didelta carnosa

Patches of bright yellow flowers carpet the coastal dunes and sandy flats of the Namib Desert thanks to the hairy salad bush. A resilient member of the daisy family, it uses a combination of semi-succulence, rolled leaves, and silvery hairs to cope with extreme drought and salt exposure. These same adaptations allow the plant to form dense mats that stabilize the shifting sands that shape the dune ecosystem – a vital role in the harsh coastal habitat.

10/ Fog potato
Solanum montanum

An unassuming cousin of the tomato and potato, the papa silvestre found in Chile and Peru is a master of desert timing. It survives for months, or even years, as a dormant potatolike tuber buried in the arid soil, waiting for optimum conditions. When the life-giving coastal fogs roll in, the plant responds with explosive growth, rapidly covering the desert in a carpet of green vegetation and delicate white or purple flowers that nod in the breeze. It completes its entire reproductive cycle in the brief window of moisture provided by the fog, before retreating back into its dormant state.

11/ Succulent Amaranth
Arthraerua leubnitziae

Succulence is such an effective way to cope in arid environments that it has evolved multiple times independently in different plant groups. Found across the Namib fog-zone and on the Skeleton Coast, the pencil bush looks like many other cactuslike succulents but belongs to the Amaranth family, in which it is cousin to beets, chard, and quinoa.

12/ Carrion flower
Hoodia currorii

The enigmatic ghaap, found deep in the Namib Desert, is a spiny succulent from the dogbane family that resembles a small cactus. The ghaap produces distinctive papery reddish-pink flowers – these blooms emit the powerful scent of rotten meat, an adaptation that attracts pollinating flies from afar.

Vibrant yellow trumpet flowers bloom for the fog season

8/

Leaf sheaths form a false stem

• COASTAL DESERTS •

9 /

10 /

11 /

12 /

• DESERTS •

Aloe vera
Originating on the Arabian Peninsula, its proximity to important historic trade routes meant that *Aloe vera* was taken into cultivation before other species. It has a long history as a medicinal plant.

Aloe vossii
Grass aloes have lost their succulence, so their leaves are camouflaged from herbivores in their grassland habitat.

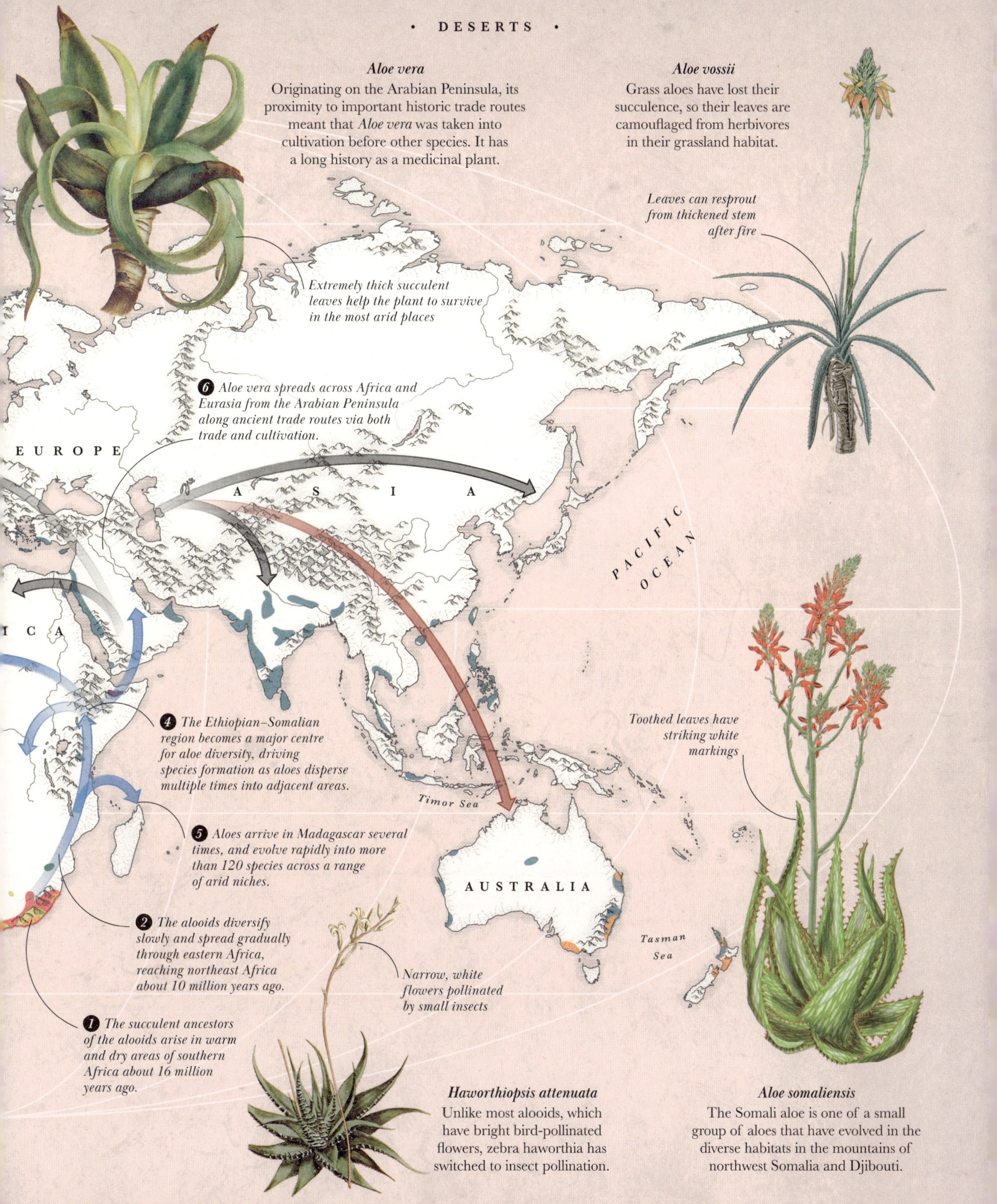

Leaves can resprout from thickened stem after fire

Extremely thick succulent leaves help the plant to survive in the most arid places

❻ *Aloe vera* spreads across Africa and Eurasia from the Arabian Peninsula along ancient trade routes via both trade and cultivation.

EUROPE

ASIA

PACIFIC OCEAN

❹ The Ethiopian–Somalian region becomes a major centre for aloe diversity, driving species formation as aloes disperse multiple times into adjacent areas.

Toothed leaves have striking white markings

Timor Sea

❺ Aloes arrive in Madagascar several times, and evolve rapidly into more than 120 species across a range of arid niches.

AUSTRALIA

❷ The alooids diversify slowly and spread gradually through eastern Africa, reaching northeast Africa about 10 million years ago.

Tasman Sea

Narrow, white flowers pollinated by small insects

❶ The succulent ancestors of the alooids arise in warm and dry areas of southern Africa about 16 million years ago.

Haworthiopsis attenuata
Unlike most alooids, which have bright bird-pollinated flowers, zebra haworthia has switched to insect pollination.

Aloe somaliensis
The Somali aloe is one of a small group of aloes that have evolved in the diverse habitats in the mountains of northwest Somalia and Djibouti.

PLANTS OF
COLD DESERTS

Cold, arid deserts exist at high latitudes and altitudes across the world's continents, far from the moderating influence of tropical zones. These harsh environments are shaped by extreme dryness and dramatic temperature swings – scorching days give way to freezing nights, while winters bring hard frosts. They may receive seasonal rainfall, but skies are usually cloudless for months on end. In North America, winter frosts distinguish cold deserts, such as the Great Basin, from hotter deserts in the south. In Asia, cold deserts such as the Gobi Desert owe their existence to the Himalayas, which form a rain shadow, blocking moisture-bearing winds from reaching these regions.

Shimmering petals reflect light to draw in pollinators

1/

1/ Underground treasure
Tulipa kolpakowskiana
In the cold deserts of Central Asia, with long freezing winters, scorching summers, and brief seasonal rains, many plants survive as geophytes – storing energy in underground bulbs or tubers. Kolpakowsky's tulip exemplifies this strategy. Emerging from dormant bulbs during fleeting spring rains, it races to bloom before drought returns. Its petals shimmer with subtle iridescence, creating an eye-catching display that helps it compete for pollinators among the desert's short-lived spring wildflower show.

2/ Pungent defence
Ferula assa-foetida
In the semi-arid regions of southern Iran, stinking gum deploys a powerful chemical defence. This drought-tolerant member of the carrot family oozes pungent latex from its thick taproot when damaged – a foul-smelling deterrent that repels predatory herbivores and insects. The smell derives from volatile sulphur compounds that react when the latex is exposed to air, moisture, or heat.

• COLD DESERTS •

2 /

Small yellow-green flowers grow on large umbels

3 /

5 /

6a /

4 /

3/ Ancient survivor
Ephedra ochreata

Members of the genus *Ephedra* are among the most ancient of seed plants and are related to *Welwitschia*, an unusual coastal desert plant with two long, ribbonlike leaves. Adapted to extreme aridity, ephedras have cut water loss to a minimum by reducing their leaves to tiny, dry scales. Photosynthesis takes place in the jointed stems, which continue to function through long dry seasons. Found in deserts in South America, *Ephedra ochreata* is one of the toughest of the group, thriving where few other seed plants can survive.

4/ Hidden waters
Populus euphratica

In autumn, among the sand dunes of the Gobi Desert of Inner Mongolia, miles from open water, lines of golden-leaved Euphrates poplar grow directly from the dunes. These trees follow underground rivers and seasonal streams, their extraordinary root systems plunging up to 20 m (66 ft) deep to tap hidden aquifers that other plants cannot reach. This remarkable adaptation allows them to flourish in extreme desert conditions where surface water vanishes.

5/ Windswept wanderer
Leontice incerta

In the windblown deserts of Central Asia, the pollinated flowers of *Leontice incerta* – a member of the barberry family – develop at least two tiny seeds, each enclosed in an extraordinary inflated bladder. As the seeds mature, these bladders detach from the mother plant. Like miniature tumbleweeds, they roll and bounce across the harsh landscape, carried by powerful desert winds. This clever dispersal strategy allows the plant to colonize distant suitable habitats, escaping competition and finding rare pockets of moisture in the vast arid steppes.

6/ Stealing to survive
Prosopanche americana

Mostly found in arid parts of South and Central America, prosopanche are extraordinary subterranean holoparasites – parasitic plants that give up photosynthesis entirely. By stealing water and nutrients from established plants, they avoid the challenge of extracting scarce resources from dry soils. Almost unrecognizable as flowering plants, they have no leaves or roots. Only the flowers are visible, looking more like alien fungi bursting from the ground. *Prosopanche americana*, one of the most widespread species, parasitizes multiple species of the bean family.

Parasitic structure
When they contact the roots of a host plant, the rounded nodules (tubercles) on the prosopanche's rhizome can siphon off water and nutrients.

6b /

Fleshy, funguslike flowers attract and briefly trap beetles to enable pollination

Some tubercles develop into sturdy flower stalks that break through the soil surface

7 / Giants of the cold
Leucostele atacamensis

The cardón cactus is well adapted to the bitter cold of Isla Incahuasi – an island on the vast Salar de Uyuni salt flats located more than 3,600 m (11,800 ft) above sea level. The cactus is covered in a thick coat of insulating hairs and it conserves its limited resources through extremely slow growth. It can take many years to reach imposing heights of up to 10 m (33 ft). Forming a columnar shape reduces the plant's surface area, which helps to limit moisture loss in the windswept environment.

8 / Desert anchor
Haloxylon ammodendron

A specialist that grows only in sandy deserts and dunes across Central Asia, saksaul secures itself in shifting sands with an extraordinary root system that provides the plant with both water and stability. Its taproot plunges to at least 10 m (33 ft) underground while lateral roots spread horizontally, providing further support. The plant's tiny scalelike leaves minimize water loss, while its green bark takes over food production and performs photosynthesis when leaves are shed.

9 / Custom-fit flowers
Calceolaria polyrhiza

The Patagonian slipper flower is found across a huge range of habitats in the cold, windswept interior of Argentina, ranging from wet bogs to dry, stony steppe. Its compact, low-growing form and tough, small leaves help it to survive the region's harsh winters and desiccating winds. It produces a variety of flower types that suit different-sized oil-collecting bee species. This pollinator-driven specialization suggests that *Calceolaria polyrhiza* may be diverging into several new plant species.

10 / Gritty deterrent
Salpiglossis spinescens

Campanula de Chañarcillo is a tough little plant known for its delicate, white, bell-shaped flowers. It is found only in the hyper arid Atacama Desert, where little else grows. Dust and sand that is blown across the desert bind to sticky hairs that cover its leaves and stems, and deliver a mouthful of grit to animals that attempt to eat the plant. The compact and dense tangle of branches also helps to reduce water loss by trapping a boundary of still air within the plant.

11 / Paper-spined cactus
Tephrocactus articulatus

In the cold deserts of central Argentina, the paper-spined cholla displays a novel adaptation. A relative of the prickly pear, it forgoes the typical sharp spines for clusters that are flat and parchmentlike. The reason for this is unclear, but these papery spines may supply insulation by trapping pockets of warmer air rather than defence against herbivores. Its stem segments are weakly attached and easily break off. This characteristic allows wind to dislodge and transport the sections across stony ground, enabling new plants to establish in an environment where survival is difficult.

7 /

• COLD DESERTS •

8 /

9 /

10 /

11 /

Chapter 4

MOUNTAINS

Snow cover

A blanket of snow can provide plants with a protective cover against frost and is often the only source of moisture in high mountains. Some plants withstand the weight of snow by forming dense cushion shapes that spread the load so that their stems do not snap.

KEY:
- Low snow cover
- High snow cover

The Rockies are home to one of the most diverse ranges of conifers in North America

Around 15 per cent of the world's plant species are found in the high-altitude tropical Andes

MOUNTAINS

High mountain altitudes create climates with strong temperature fluctuations and variable precipitation. Snow and ice persist for most of the year, and with little or no soil, intense solar radiation, strong winds, and a short growing season, mountain plants face a daunting challenge.

ADAPTING TO ALTITUDE

Low mountains often feature abundant tree cover that progressively changes in character, transitioning from deciduous broadleaves to conifer forests as the elevation increases. On high mountains, a boundary forms – the tree line – above which conditions are too harsh for most trees. Exposed areas above the tree line are dominated by low-growing shrubs and grassland.

Climate highs and lows

On tall mountains, precipitation is often high and temperatures are much cooler; atmospheric pressure declines as altitude increases, and results in lower temperatures. Low mountains vary less in temperature but precipitation is highly dependent on the seasons.

KEY
- Páramo
- High mountains
- Low mountains
- Average temperature

MOUNTAINS

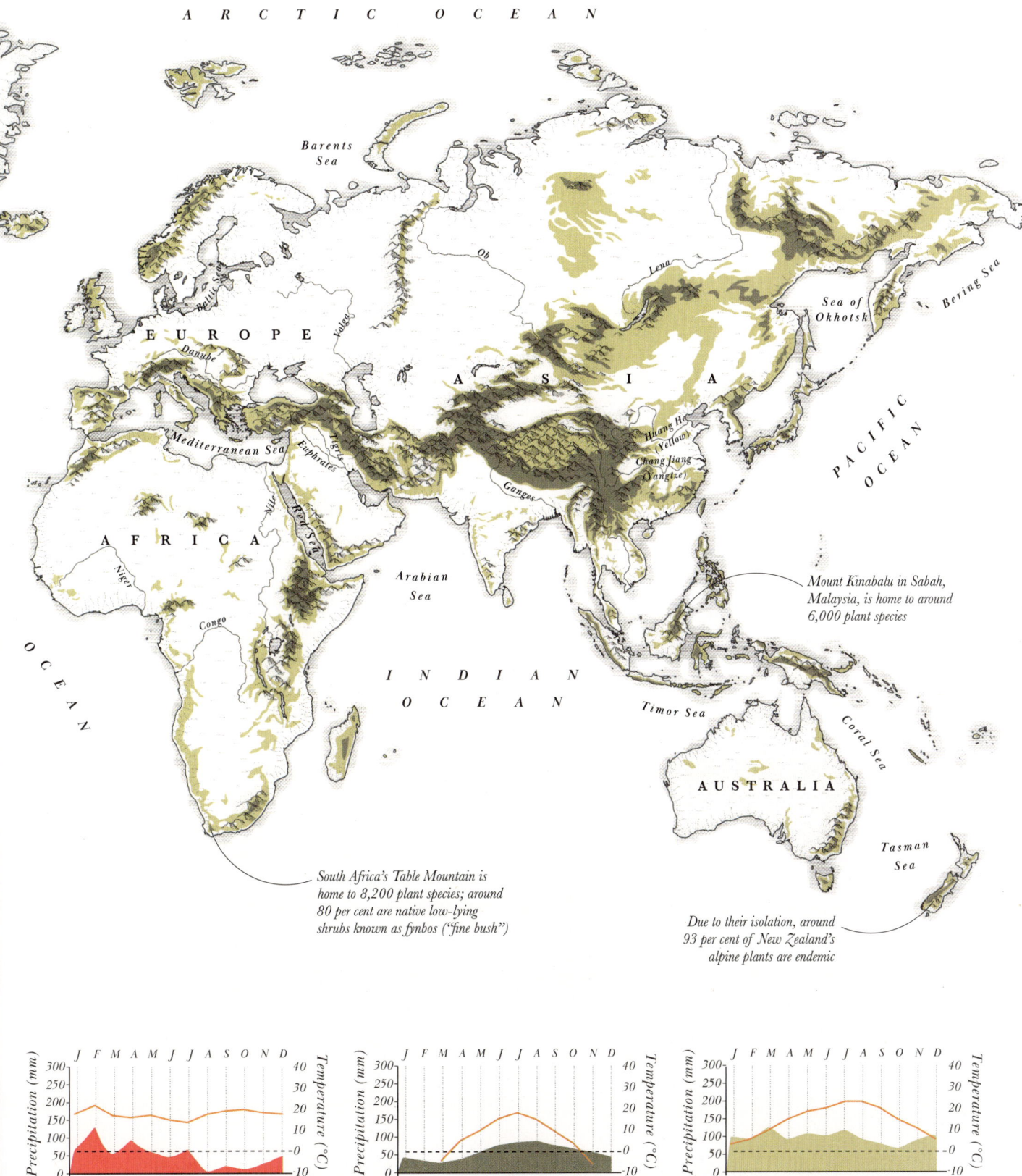

Mount Kinabalu in Sabah, Malaysia, is home to around 6,000 plant species

South Africa's Table Mountain is home to 8,200 plant species; around 80 per cent are native low-lying shrubs known as fynbos ("fine bush")

Due to their isolation, around 93 per cent of New Zealand's alpine plants are endemic

PÁRAMO

HIGH MOUNTAINS

LOW MOUNTAINS

ADAPTATIONS

ALTITUDE AND EXPOSURE

Mountain environments present plants with a particular combination of challenging conditions. Life becomes more difficult as altitude increases, leading to zones of vegetation that can appear more and more specialized with higher elevations. High mountain areas can be very arid – water may come directly from clouds or snow melt rather than from rain. The air can be dry as winds strip away moisture, causing plants to develop particular architectural growth forms to cope with life on exposed sites.

fig. 1 Laburnum alpinum; fig. 2 Picea abies; fig. 3 Empetrum nigrum subsp. nigrum; fig. 4 Ranunculus acris; Centaurea nigra; fig. 5 Poa alpina; fig. 6 Silene acaulis; fig. 7 Polytrichastrum alpinum; fig. 8 Papaver alpinum; fig. 9 Oreopolus glacialis; fig. 10 Echeveria quitensis; fig. 11 Diapensia lapponica; fig. 12 Echium wildpretii; fig. 13 Argyroxiphium sandwicense; fig. 14 Dendrosenecio kilimanjari

fig. 9

Densely packed stems form mounds that trap warmth

Cushion

fig. 10

In arid rock crevices, rosettes channel water to the base of the plant

Succulent rosette

Dish-shaped flowers focus warmth onto the stamens and ovaries

Low rosette leaves may be up to 20 °C (70 °F) warmer than the surrounding air

fig. 8

Herbaceous rosette

Growth structure

A plant's architecture can be crucial to its survival in alpine areas. Many high mountain plants stay close to the ground, but adopt a range of different growth habits, such as cushion shapes or rosettes.

High altitude ←

fig. 7

Moss forms low growing, drought tolerant colonies

MOSSES AND LICHENS

fig. 6

Small flowers cover the surface of the plant

CUSHION PLANTS

fig. 5

Grasses begin to dominate at exposed higher altitudes

GRASSLAND

fig. 4

A profusion of flowers attracts seasonal pollinators

ALPINE MEADOW

MOUNTAINS

Coping with exposure

Ultraviolet (UV) light can damage DNA, so plants at the highest altitudes must protect themselves. They use an array of structures and coatings to do this – often giving the plant a grey sheen.

fig. 11
A cuticle covered in wax reflects UV rays

Waxy cuticles

fig. 12
Silvery hairs cover all exposed parts of the plant
Silvery hairs reflect UV light and help to reduce water loss

Silver hairs

Surviving cold

Protection against the cold is essential for plants. It can be disastrous if their cells freeze because this destroys the potential for regrowth. Plants have come up with a number of insulating strategies to protect their growing points.

fig. 13
Thick, woolly hairs keep cells warm

Insulation hairs

fig. 14
A ruff of dead leaves insulates the main crown
Withered foliage insulates the stem

Leaf ruff

Vegetation zones

Mountain vegetation zones vary in different latitudes, as well as on the sunny and shady sides of an alpine site. On slopes that face away from the sun, each zone tends to finish at a lower altitude.

fig. 1
BROADLEAF TREES
Broadleaf trees are mostly deciduous in temperate areas
Flowers at low altitudes attract plenty of pollinators

fig. 2
NEEDLE TREES
Needles help to prevent water loss
Evergreen habit allows the tree to maximize the growing season in cold, high altitudes

fig. 3
SHRUBS
Shrubs become lower growing as elevation increases

Low altitude →

ADAPTATIONS

ISOLATION AND TEMPERATURE

Mountains are effectively islands – isolated from neighbouring peaks in a "sea" of lowlands. Each has a distinct ecology with specific mixtures of plant and animal species that rely on – and exploit – each other, but plant pollinators may be scarce and scattered. At high altitudes, plants must cope with ground temperatures that are significantly warmer than air temperatures, as well as the extreme variation between day and night. Loose, stony soils are often poor in nutrients, but plants have found ways to access and store the resources that they need, and anchor themselves in shifting substrates.

fig. 1 Kalmia latifolia; *fig. 2* Meconopsis punicea; *fig. 3* Pulsatilla alpina; *fig. 4* Meconopsis betonicifolia; *fig. 5* Primula minima;
fig. 6 Sempervivum montanum; *fig. 7* Noccaea caerulescens; *fig. 8* Onobrychis cornuta; *fig. 9* Ebenus cretica; *fig. 10* Armeria maritima

fig. 1
Spring-loaded stamens catapult pollen onto visiting insects

fig. 3
Open flowers can accommodate a wide range of visitors

Ease of access

fig. 2
Large, bright nodding flowers provide insects with shelter

Nodding shelters

Trigger mechanisms

Maximizing pollination
Wind pollination may work for dense stands of grasses in alpine meadows, but unpredictable high winds are less reliable at delivering pollen to scattered wildflowers. Instead they depend on pollinating insects, and use many strategies to entice them.

fig. 4
Robust flower heads elevate the blooms to attract distant pollinators

Flowers held high

• MOUNTAINS •

Temperature variation

The temperature differences in mountain habitats can be dramatic – from the base to the peak, during night and day, and even between the air and the sheltered parts of a plant – a phenomenon that many species attempt to exploit.

Cushion plants retain warmth through their dense growth form

Centres of succulent rosettes absorb and retain warmth

Cosy cushions   *fig. 5*   *fig. 6*   Warm hearts

Toxic nickel and zinc are stored in shoots

Fruits fall near the base of the plant to ensure that their offspring germinate in favourable limestone soils

fig. 7

Absorbing toxins

Toxic metals can be excreted through glands on the leaves

A deep taproot anchors the plant in rock crevices

fig. 9

Soil specialists

fig. 8

Firm anchors

Tough terrain

Specialized rock-dwelling lithophytes find a foothold in crevices and shifting scree slopes with little nutrient-rich organic matter. Some plants have adapted to specific rock types – such as calcicole plants, which thrive on limestone.

fig. 10

Discarding toxins

Narrow, threadlike petals with stamens and anthers between them, surrounded by a variable number of greenish sepals

1 /

1 / Protecting pollen
Paris polyphylla
Found at altitudes up to 2,500 m (8,200 ft), this shade-loving plant thrives in the understorey of broadleaf forests in China. It forms stems with a single whorl of 5 to 20 leaves and bears flowers with a special talent for protection. At night and when it rains, the flower anthers close to protect the pollen within from cold and moisture, safeguarding reproduction.

PLANTS OF
LOW MOUNTAINS

Low mountains are categorized as the slopes of any mountains up to the limit of the tree line, below the level of alpine meadows and shrublands. The tree line varies according to latitude, ranging from a few hundred metres in the Scottish Highlands to above 4,000 m (13,120 ft) in tropical Southeast Asia. Low mountains are characterized by various types of forest, depending on the elevation and orientation of the slopes. For example, slopes facing away from the prevailing rain-bearing winds, will be in a rain shadow, with little precipitation. As slopes rise, changes in conditions and soil composition also affect vegetation, creating distinct growing zones at different altitudes.

2/ Butterfly magnet
Buddleja davidii
Originating from the mountain slopes and rocky stream banks of central China, *Buddleja* is cultivated and naturalized worldwide. It has a high tolerance for poor, calcium-rich, disturbed soils and its tough root system allows it to anchor itself in unstable rocky ground and shingle. A fragrant source of nectar, it attracts butterflies and moths to aid pollination.

3/ Synchronized flowering
Fargesia nitida
Dense forests of this clumping bamboo are found in the mountains of China. It exhibits mass blooming, in which all the offspring of a single plant flower at the same time, dying en masse afterwards. This phenomenon happens once every century and is believed to maximize pollination, mitigate seed predation, and reduce competition for new seedlings. However, forests experiencing this synchronicity take several years to regenerate, causing food shortages for the giant pandas who rely on the bamboo.

4/ Size wise
Akebia quinata
This vine from East Asia produces separate pink-purple male and female flowers on the same plant. A few large female flowers open near the base of the flower cluster, while several smaller male flowers appear at the tip. By virtue of their larger size, the female blooms attract pollinators first; any pollen these visitors may be carrying from other flowers is deposited here before they move on to the male flowers above, thus avoiding self-pollination.

5/ Back to life
Ramonda myconi
A species endemic to the Mediterranean, this long-lived plant grows in shaded rock crevices of the Pyrenees mountains. Known as a "resurrection plant", the Pyrenean violet is one of the few plants that can revive when it is rehydrated after having withered and wilted in dry conditions. It survives the bitter cold of winter by concentrating certain sugars in at-risk cells, stabilizing its tissues.

6/ Altitudinal adaptation
Rhododendron arboreum
The shape, form, and size of this plant varies depending on its environment. Shrublike in exposed areas with limited sunlight, it can grow to a towering tree 15–20 m (49–66 ft) tall, when conditions are more favourable. The colour and nectar guide patterns of its flowers differ at different altitudes, to ensure it attracts the available pollinators.

Thick, leathery leaves withstand wind and cold, and reduce water loss

7 /

9 /

8 /

10 /

11 /

• LOW MOUNTAINS •

7/ Showy display
Lewisia cotyledon
This evergreen grows in rocky, subalpine sites in the mountains of western North America. From spring to late summer, and sometimes into autumn, a succession of tall stems carry showy flowers above low rosettes of fleshy leaves. The long blooming season and the flowers' vibrant and varied colours, which help them stand out against their rocky habitat, increase the chances of attracting pollinators in the plant's remote, rugged environment.

8/ Pollen deluge
Schizanthus hookeri
This plant from the foothills of the Andes is known as the poor-man's orchid for its orchidlike flower. The flower has an explosive pollination mechanism, similar to pea flowers, whereby the largest petal acts as a banner to attract bees; when a bee lands on the keel, it triggers the release of stamens that are held under tension. The stamens dust the bees with pollen to carry to the next flower for cross pollination.

9/ Bright and beckoning
Paeonia daurica
Known as Molly the witch – perhaps the result of a mispronunciation of the original species name *P. mlokosewitchii* – this perennial plant occurs in the Caucasus region and favours shady broadleaf forests. It develops large capsules that open to expose shiny black, fertile seeds interspersed with smaller, fleshy, red unfertilized ovules that serve to attract potential seed dispersers such as birds and mice.

10/ Back-up plans
Cardiocrinum giganteum
The giant lily grows up to 3 m (10 ft) tall and is native to woodland clearings in the Himalayas. It has a monocarpic lifecycle, meaning it flowers only once, then dies; it has various strategies, however, to ensure that life goes on. It produces clones in the form of tiny bulblets at the base of its stem, which break off to form new plants. Its flowers can also self-fertilize with their own pollen and set seed without the help of visiting pollinators.

11/ Famine and feast
Cedrus deodara
Found in the foothills of the western Himalayan region between 1,200–4,000 m (3,940–13,120 ft), this tree occasionally forms single species forests. It is adapted to the extremes of drought and monsoon by having a deep root system that can access water in dry spells, while providing an anchor during heavy rains. Its waxy, needlelike leaves also prevent excess water absorption and rain damage in wet spells, but reduce water loss in dry conditions.

12/ Strength in numbers
Campanula glomerata
This hardy perennial is found in temperate parts of Eurasia and can withstand temperatures as low as −20°C (−4°F). It produces dense clusters of purple flowers, which mature into seed capsules. This profusion of nectar-rich blooms attract many pollinators, which, in turn, leads to a substantial seed set. Released from the capsules by the wind, most seeds fall and take root near the parent plant.

Style protrudes from bell-shaped flower to facilitate insect pollination

12a /

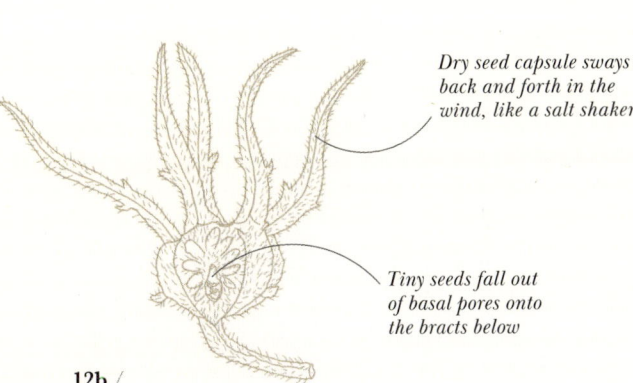

Dry seed capsule sways back and forth in the wind, like a salt shaker

Tiny seeds fall out of basal pores onto the bracts below

12b /

Spreading seeds like salt
Each fertilized campanula flower matures into a porous capsule containing numerous tiny seeds. These seeds are released onto the bracts below to be spread in a portioned way.

13/ Aggressive colonizer
Impatiens glandulifera

Often found growing near bodies of water, Himalayan balsam is a 1–2 m (3–7 ft) tall annual. It is native to mountains in Pakistan and northwest India, but has become naturalized in Europe. It outcompetes plants in wet sites by producing abundant quantities of nectar, to attract pollinators, and by having exploding seed pods that can disperse seeds up to 7 m (23 ft) away. The seeds are carried further still on the waters of nearby rivers.

14/ Deceptive scent
Magnolia sprengeri

This medium-sized tree grows on the mountain slopes of central China. Its large flowers, pollinated mainly by beetles, have a female phase that does not provide the pollinator with a reward, followed by a male phase with pollen as a food reward. At the peak of each phase, the flowers produce heat and an attractive scent. This fools the beetles into visiting a flower during both sexual stages – first fertilizing the flower with pollen it has carried from other blooms (for no reward); then picking up its pollen (reward) for transfer to the next flower it goes on to visit.

15/ Erosion control
Pinus mugo

The dwarf mountain pine grows at the tree line in the mountains of central and southeastern Europe. A dense, low-growing, and spreading plant, it plays an important role in slowing down avalanches, preventing soil erosion, and promoting soil formation. It does this by having deep, widespread roots that bind soil particles together, and a dense canopy that shields the soil surface from splash erosion. Its fallen needles also create a thick layer of leaf litter, providing a nourishing soil mulch.

16/ Close dependents
Roscoea purpurea

Found in the colder mountain regions of the Himalayas, this plant has coevolved with the long-tongued fly, and the two species depend on each other for survival. When the fly probes one of the plant's flowers for nectar, it pushes against a hingelike appendage that lowers the anther. Since its male and female organs do not mature at the same time, either the anther deposits pollen on the fly's back, or the stigma receives pollen that the fly is already carrying.

17/ Tailored access
Aquilegia formosa

Columbines are most diverse in the Rocky Mountains of western North America, where the western columbine is found. Its flowers have very long spurs, adapted to ensure that nectar can only be accessed by their long-tongued pollinators, hummingbirds. This natural filter reduces visits from short-tongued non-pollinating insects, who might steal the valuable nectar without transferring any pollen.

18/ Seeking warmth
Echium wildpretii

This subalpine biennial is native only to the dry volcanic slopes of Tenerife and La Palma on the Canary Islands. Known as "tower of jewels" for its conical habit, it comprises a basal rosette of leaves, from which rises a tall panicle of tightly packed red flowers. It is intolerant of low temperatures and thrives only on sites with many hours of strong sunlight, low rainfall, and excellent drainage. This narrow climatic niche has limited its natural distribution to a very small area.

19/ Potent defence
Aconitum napellus

Native to the mountainous regions of Europe, monkshood is considered one of the most poisonous plants in Europe, due to the toxic alkaloids in its roots, stems, leaves, flowers, and seeds. This defence system is designed to deter browsing herbivores and insect thieves, who might steal nectar without pollinating the flowers. Long-tongued bumblebees, the plant's true pollinators, are more resistant to the toxins.

13 /

Spurs are tubular extensions containing nectar

• LOW MOUNTAINS •

14 /

15 /

16 /

17 /

18 /

19 /

Bergenia purpurascens
Purple bergenia is a perennial native to the mountainous regions of East Asia. It grows in subalpine habitats with large leaves held close to the ground that turn red with the cold.

Purple-red, drooping flowers are bell-shaped

Small flowers appear on short stems close to the ground

③ Saxifraga diversifies in North America and spreads further along the spine of the Andes down to Tierra del Fuego.

SAXIFRAGES

The saxifrage family (Saxifragaceae) arose around 38 million years ago as herbaceous plants and diversified – adapting to wet environments and temperate forests, as well as alpine and Arctic habitats.

Originating in northwest North America, saxifrages spread across the Northern Hemisphere as a result of migration, diversification, and hybridization. Comprised mostly of perennial plants, the family includes well-known ornamentals, such as *Bergenia*. Its five-petalled flowers may be bell- or star-shaped, and the leaves often appear in basal rosettes.

HISTORIC MIGRATION
Saxifrage →

REPRESENTATIVE LIVING SPECIES
- Bergenia purpurascens
- Darmera peltata
- Lithophragma parviflorum
- Saxifraga bicuspidata
- Saxifraga oppositifolia
- Saxifraga stolonifera

Flowers have two elongated petals to attract pollinators

Saxifraga stolonifera
Creeping saxifrage grows in moist forest areas of South Asia and is naturalized worldwide. It spreads by producing long stolons that develop plantlets at the ends.

PLANTS OF
HIGH ALPINE AREAS

Found all over the world, including the Alps, Andes, and Himalayas, high alpine mountain areas are generally above 3,000m (about 10,000ft) and beyond the tree line, but the altitude depends strongly on the distance from the equator. The sub-biome is characterized by extreme temperatures, a short growing season, high sun radiation, low precipitation often in the form of snow, and desiccating windy conditions. Plants are adapted to survival by cushionlike growth forms, water-retaining leaves, and an extensive root system. Additionally, the formation of mountains creates a mosaic of soil types that require plants to make specific adaptations to absorb nutrients.

1/ Isolated communities
Dryas octopetala
An Arctic-alpine flowering plant in the rose family, *Dryas* is found in Arctic regions, Europe, and North America. The name refers to the eight petals in the flower, an unusual number in a family where five is most common. The plant is a typical example of a species that has formed in geographically separate areas, in this case caused by the retreat of the ice sheet, leaving isolated populations in more exposed or high-altitude environments.

2/ Evolution in action
Gymnadenia austriaca
This terrestrial orchid is endemic to the European Alps and Pyrenees, where it is mostly found in high-altitude grassland. Its flowers have a shorter lip (labellum) than most other orchids and it has a less-developed spur than related species. These traits – examples of an adult plant retaining its juvenile form – could be an evolutionary adaptation caused by a change in the behaviour of its pollinators.

3/ Wrapped up in wool
Raoulia eximia
Also known as vegetable sheep, *Raoulia* grows in the high-altitude Southern Alps of New Zealand. It is characterized by a compact cushion-like growth form with only the growing tips visible, which resembles sheep from a distance. The stems are extensively branched and bear small leaves, tightly packed at the extremity of the twigs, forming a solid mat. This acts as insulation against cold wind and intense sunshine, minimizing water loss.

4/ Lateral thinking
Lobelia deckenii
This species grows in the alpine zone of Mount Kilimanjaro in East Africa, producing giant rosettes of leaves with huge spikes of flowers. Flowering stems die after blooming, but the plant survives by producing lateral rosettes. Water stored in the rosettes freezes into crescent-shaped ice formations at night to protect the growing point of the shoot. This has led to its nickname "gin and tonic lobelia".

Eight petals surround the many yellow stamens

HIGH ALPINE AREAS

5/ **Tiny meadow flower**
Gentiana verna
One of the smallest gentians, spring gentian has a widespread distribution in the Arctic-alpine meadows of Eurasia and is even found in Britain and Ireland as a surviving remnant of a population that was formerly more widespread. It grows in sunny, exposed grassland and occurs in alkaline and acidic soils. At higher altitudes leaves are generally smaller than at lower altitudes because of the harsher growing conditions. However, although the British Isles populations grow at lower altitudes, they resemble the smaller-leaved high-altitude populations on continental Europe, which may be the result of genetic drift.

5/

Prominent central vein

Striking deep-blue flowers unfold from pyramidlike buds

6 /

7 /

8 /

9 /

10 /

11 /

6/ Supercool tree
Polylepis tarapacana

This short tree is distributed in volcanic soils in the high Andes up to an altitude of 5,200 m (17,000 ft), forming the highest altitude forests in the world. *Polylepis tarapacana* is well adapted to the harsh, dry conditions with its thick, multi-layered bark. It also avoids freezing through a clever process called supercooling, by increasing its sugar content to lower the freezing point of water inside the tree. Its flowers lack petals and are wind-pollinated, which neatly sidesteps the scarcity of pollinators at the altitude it inhabits.

7/ Fake nectaries
Parnassia palustris

Enjoying an extensive range across temperate and subarctic regions in the Northern Hemisphere, grass-of-Parnassus lives in wet environments such as bogs and wet alpine marshland. Its veined cup-shaped flowers contain a cluster of five sterile stamens, which branch into multiple fingerlike lobes ending in a glandular knob. Captured by sunlight, the glistening droplets mimic nectar secretion and act as a cue for insect pollinators. Their real reward is a limited amount of nectar secreted on the common base of the stamens.

8/ Adapted to extremes
Meconopsis horridula

The prickly blue poppy is a high-altitude Himalayan species that is typically covered with spines on its leaves and stem. Plants grow in exposed, cold conditions and contain high levels of flavonoids that protect against the high UV radiation. Transplanting experiments show that it is well adapted to photosynthesize at lower temperatures, but performs poorly at lower altitudes.

9/ Asexual alternative
Bistorta vivipara

The alpine bistort can be found in mountainous areas across Europe, Asia, and North America. The lower flowers on its inflorescences are replaced by small bulblike structures called bulbils, which can grow into new plants. As the alpine bistort inhabits harsh environments with a lack of pollinators, it rarely produces viable seeds, so its method of asexual reproduction by bulbils is a viable alternative for the species.

10/ Animal magnetism
Acaena alpina

This widespread subshrub is found along the Chilean and Argentinian Andes. Flowers are produced at the very edge of the stem and become burrs that cling to the fur of animals. Its fruits are dispersed by the development of numerous barbed spines on the inferior ovary that also attach to passing animals (a method of dispersal known as epizoochory).

11/ Pecking order
Calceolaria uniflora

The genus name of this unusual plant native to southern Patagonia and Tierra del Fuego comes from the Latin "calceolus", meaning "little shoe", in reference to the shape of the flowers. *Calceolaria uniflora* produces specialized pouch-like petal tubes consisting of two lips. Flowers are pollinated by a bird, the least seedsnipe (*Thinocorus rumicivorus*), and pollen is transferred from the two lateral stamens when it pecks at the conspicuous white appendages on the lower lip, which are rich in sugar.

12/ Woolly sun block
Leontopodium nivale

Edelweiss is found in the mountains of Europe stretching from the Pyrenees to the Balkans, growing on limestone in rocky crevices. What appears to be flowers are 2–10 yellow flower heads that are surrounded by a rosette of bractlike woolly-white leaves that reflect UV radiation. The grey-white covering is most obvious in plants growing at high altitude and disappears when grown in lowland gardens where plants are taller with fewer hairs.

12 /

Woolly-white leaves form a star shape

13/ Fringed blooms
Soldanella villosa

This clump-forming, evergreen alpine perennial from the primrose family is native to the Pyrenees, hence its common name of Pyrenean snowbell. Its flowers are bell-shaped with purple, fringed petals appearing on hairy stems. The flowers are thought to produce a whorl of sterile stamens that become incorporated in the petals (corolla) as they develop.

14/ Alpine greenhouse
Rheum nobile

This giant rhubarb is native to the Himalayas and produces erect, conical flowering stems up to 2 m (6.5 ft) covered by cream-coloured regularly overlapping bracts. The large translucent bracts create a glasshouse effect by sheltering the flowers and increasing the temperature while protecting them against the intensity of UV radiation. This allows the species to thrive in high alpine areas as one of the largest high-altitude perennials.

15/ Underground larder
Incarvillea delavayi

Occurring in the high-altitude regions of Yunnan in China, herbaceous *Incarvillea delavayi* is unlike other members of the Bignoniaceae family – which are tropical and subtropical trees or climbers. It is able to grow in exposed high mountain meadows with moist, well-drained soils by producing large tuberous roots. The root system consists of a thickened taproot functioning as underground storage for the extreme winter conditions.

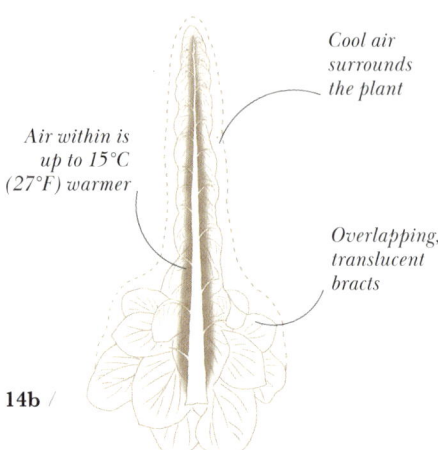

14b /

Natural greenhouse
The translucent bracts of this ingenious rhubarb protect it from the extreme low temperatures of the Himalayas by trapping warm air around the developing flower.

16/ Protective coat
Saussurea gossypiphora

The woolly snowball is found in the high Himalayas and belongs to the aster family. The species is well adapted to the high mountains as it has developed long white hairs that cover the stem, leaves, and large bracts that enclose the flowers. Flower heads with purple florets are embedded in woolly hairs, which provide insulation and protection against frost, UV radiation, and desiccation by wind.

17/ Specialized water storage
Sempervivum montanum

The mountain houseleek occurs in the mountains of southern and central Europe, thriving in arid, rocky siliceous soils with excellent drainage. The succulent fleshy leaves are well adapted to store water and a specialized photosynthetic pathway allows plants to absorb carbon dioxide by opening tiny pores (stomata) on the surface at night to avoid water loss during the day.

18/ Stabilizing the soil
Saxifraga oppositifolia

The purple saxifrage is widespread in the alpine-Arctic environments of Europe and North America. Its low, cushionlike shape, which helps it to conserve heat, is one of the multiple factors that makes it well adapted to grow in high-altitude mountain screes where the rock formations are loose and prone to movement from snow and alternating cycles of melting and freezing.

13 /

• HIGH ALPINE AREAS •

14a /

16 /

17 /

15 /

18 /

Yellow sunflowerlike blooms attract bee and beetle pollinators

"The impression of being confronted with a different world."

José Cuatrecasas, *A Systematic Study of the Subtribe Espeletiinae*, 2013

Hairs on leaves provide insulation and protection from UV rays

Central pith accumulates water from mist

Frailejónes
Espeletia spp.

An endemic plant of the Andean Páramo grasslands, *Espeletia* includes some of the largest rosette-forming species in the Asteraceae or daisy family; some individuals can grow up to 7 m (23 ft) tall. The common name for the genus, frailejónes, translates to "big friar" because in the mists of the high mountains these towering plants can look like robed people. This misconception reputedly thwarted an invasion of the Colombian Andes by Spanish forces in 1536. Conquistador Nikolaus Federmann's troops are said to have retreated in fear after mistaking the plants silhouetted along the mountainside for row upon row of enemy soldiers waiting for them in the mist.

To the inhabitants of the Páramo, the parameros, frailejónes is considered one of the region's most representative plants – and they use almost every part of it. The roots are used to ease digestion and to treat fevers, circulation problems, and coughs, while the resin is used in bookbinding and construction. Frailejónes are so strongly associated with the Páramo that a main character in the Colombian animated children's TV series *Cuentitos Mágicos* is based on the plant. The popular character protects water resources, and defends the environment – and in doing so brings greater public awareness to the fragile environment of the Páramo.

RESCUE MISSIONS

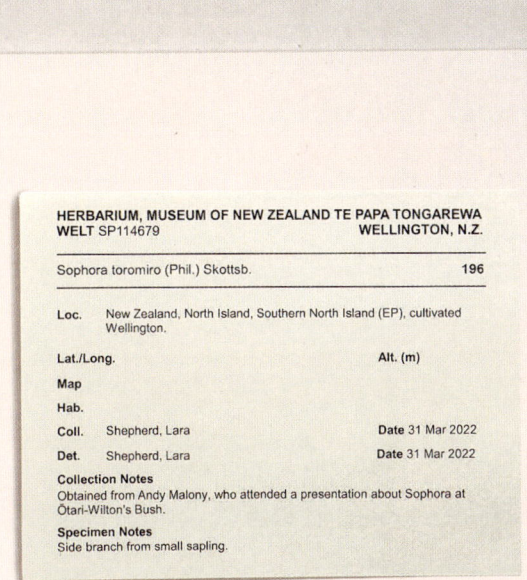

Island species are particularly vulnerable to extinction, especially if they are overexploited by humans or confronted with a newly introduced competing species. One member of the pea family, *Sophora toromiro*, is an example of a plant that has become extinct in the wild, but survives in cultivation thanks to the off-site conservation work of botanic gardens around the world.

Sophora toromiro is a small tree with bright yellow flowers from Easter Island in the South Pacific. It has cultural significance for Rapa Nui people, who used it to make sacred objects and wooden figurines in the style of the island's iconic moai stone faces. Known simply as toromiro, it became extinct in the wild around 1960 due to deforestation and overgrazing – but not all hope was lost. Seeds from the final specimen that had been taken to the Viña del Mar National Botanical Garden in Chile in the 1950s were germinated successfully and produced just under 100 trees. However, attempts to reintroduce toromiro to Easter Island have not been fruitful – possibly because the Island's population of *rhizobia* soil bacteria on which toromiro depends has also become extinct.

A ray of light has arrived in the form of a recent project led by the UK's Royal Botanic Gardens, Kew and the Gothenburg Botanical Garden in Sweden. The research focus is on identifying related *rhizobia* bacteria from Oceanian islands to grow in toromiro's root nodules. If researchers find a match, it may be possible to reintroduce toromiro to its native environment. Ultimately, although human intervention has been responsible for the extinction of toromiro in the wild, it can also be employed to resurrect a part of Easter Island's heritage.

Lost heritage
A preserved specimen of toromiro kept at the Museum of New Zealand Te Papa Tongarewa in Wellington was acquired in 1870, but only identified through DNA analysis in 2020. It may have been grown from seeds collected during Captain James Cook's historic Second Voyage, which took place between 1772 and 1775.

PLANTS OF THE
PÁRAMO

Páramos are high-elevation ecosystems found above the tree line and below permanent snow in the northern Andes, especially Colombia, Ecuador, Venezuela, and northern Peru. These "alpine islands" are shaped by intense solar radiation, thin air, strong winds, nightly frosts, and typically high humidity. The soils are acidic and waterlogged, creating bogs, lakes, and spongy ground that store vast amounts of carbon and regulate watersheds. Despite these harsh conditions, páramos support extraordinary biodiversity and endemism, making them vital reservoirs of life and essential sources of freshwater.

1/ Sky island inhabitant
Puya thomasiana
Unlike most bromeliads, which grow as epiphytes on tree branches, Puya are terrestrial. Their leaves form large rosettes topped by towering flower spikes that are visited by hummingbirds, bees, and bats. Spectacled bears feast on their nectar-rich flowers for energy. Puya grow in isolated pockets on mountain peaks. This isolation from each other results in high species diversity and endemism.

1a /

Yellow-green tubular flowers have three petals

1b /

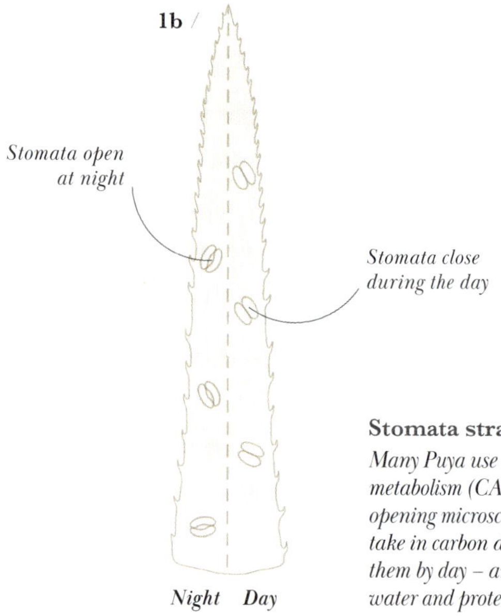

Stomata open at night

Stomata close during the day

Night   Day

Stomata strategy
Many Puya use crassulacean acid metabolism (CAM) photosynthesis, opening microscopic stomata at night to take in carbon dioxide (CO_2) and closing them by day – an adaptation that conserves water and protects against the cold.

PÁRAMO

2/ Leafy defence
Chuquiraga jussieui
Shrubs of the páramo often evolve dense, microphyllous foliage – tiny, hardened leaves covered in hairs or scales that guard them against frost, wind, and intense sun. *Chuquiraga* exemplifies this strategy with its tough, hard-pointed leaves that also act as a deterrent to most herbivores. Vivid orange flower heads at the ends of the stems blaze across the alpine landscape, providing vital nectar for hillstar hummingbirds, one of the páramo's most specialized pollinators.

3/ Bouquet plants
Chaetolepis microphylla
Some shrubs offset their tiny leaves with a profusion of flowers. Known as "bouquet plants", they rely on bright, showy blossoms to draw pollinators across the open páramo. This example produces brilliant flower clusters at the ends of small, tough-leaved stems, creating a striking floral display that transforms the stark alpine landscape into beacons for visiting insects and birds. It has smaller flowers compared to its close relatives, but still attracts pollinators through its vibrant colours.

4/ Rolling moss
Grimmia longirostris
Near glaciers, relentless freeze–thaw cycles sculpt moss into free-living spheres. These "moss balls" detach from the ground and roll with the wind or meltwater across alpine surfaces to more hospitable environments. This unique species survives this extreme life untethered, thriving as one of the páramo's most unusual moss forms.

5/ Buzz pollination
Monochaetum humboldtianum
This species from the northern Andes was named after Prussian botanist Alexander von Humboldt, whose expedition contributed to its discovery. As with most members of the Melastomataceae family, it has poricidal, or porous, anthers. The pollen held in these tubular anthers is only accessible to bees capable of shaking the pollen out of the tubes using the vibrations of their flight muscles. This deters inefficient pollinators and optimizes the pollination process.

2/

3/

4/

5/

Each stamen has two "arms"; the smaller arm is an appendage for bees to perch on when feeding

6 /

7 /

8 /

9 /

• PÁRAMO •

6/ Closed for the night
Gentiana sedifolia
Like many montane gentians, this species displays "nyctinasty", meaning that it closes its petals at night and in cloudy weather. This change can happen quickly in the páramo, where the weather changes abruptly. This adaptation prevents pollen damage and conserves energy. The plant's compact form reduces exposure to winds and its tight basal leaf rosettes trap heat to help it through nightly freezes.

7/ Forests at the edge
Polylepis spp.
Trees in this genus form the world's highest-elevation forests, marking the boundary between páramo and alpine zones. Their name means "many scales", a nod to their thick, peeling bark that protects against ground fires. With about 45 species across the Andes, these are ecologically vital, yet their range is heavily reduced. About 90 per cent of original *Polylepis* cover has been lost due to human activities.

8/ Protective tufts
Cinnagrostis recta
Bunch grasses are among the most abundant plants in the páramo, carpeting slopes with dense tufts of leaves. This growth form shelters the delicate shoot tip deep within the clump, protecting it from frost, wind, and grazing animals. By stabilizing soil and resisting extremes, bunch grasses form the resilient backbone of many high-Andean grasslands.

9/ Architects of the bogs
Sphagnum spp.
Dominating bogs worldwide, in the páramo *Sphagnum*'s waterlogged mats flourish under the cold, wet conditions. Its spongy branches draw moisture upward like capillary pumps, storing vast amounts of water. By acidifying its surroundings and locking nutrients away in its tissues, this moss shapes alpine wetlands where only specialized plants persist, making it a true ecosystem engineer of the high Andes.

10/ Avoiding the sun
Jamesonia verticalis
In the high Andes, *Jamesonia* ferns have evolved slender, upright fronds lined with tiny horizontal leaflets. This unusual form reduces exposure to intense solar radiation, echoing the tiny, tough leaves of microphyllous shrubs of the páramo. What looks like a leafy branch is actually a frond – an innovation that has evolved multiple times from sprawling, vinelike ancestors adapting to alpine conditions.

Small, leathery leaflets withstand exposed conditions

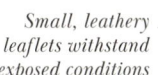

11 / Dawn blooms
Orthrosanthus chimboracensis
The morning flag is a geophyte, storing its rhizomes safely underground to protect them from harsh alpine conditions. Each day it produces delicate blue or lavender flowers that bloom briefly, often lasting only a few hours, to conserve resources. Its blossoms unfurl at dawn like a flag, offering a short-lived but striking spectacle.

12 / Icons of the Páramo
Espeletia spp.
Frailejones are iconic páramo plants, instantly recognized by their thick stems and fuzzy leaves. Species of *Espeletia* radiated rapidly across Colombia and Venezuela, developing diverse growth habits from rosettes to towering shrubs. Their erect forms and woolly foliage protect against intense sun, frost, and wind, making them paragons of high-Andean adaptation.

13 / Hugging the earth
Disterigma empetrifolium
In the páramo, this low-growing shrub survives by hugging the ground. Its prostrate branches creep along or beneath the soil surface, shielding the plant from frost and desiccating winds. Often, only the current year's shoots and leaves emerge above ground, while the rest of the shrub remains hidden, a strategy that allows it to persist in the alpine cold. It bears tiny pink urn-shaped flowers that are held upright to aid pollinators.

14 / Guiding light
Eryngium humile
This plant stands out in the páramo with its dense heads of tiny flowers surrounded by silvery bracts. These reflective bracts amplify visibility in the brilliant alpine light, serving as beacons that guide pollinators from a distance. By concentrating colour and shine around the inflorescence, this unusual strategy ensures the plant's small flowers are not overlooked in the high Andes.

15 / Mat-forming microhabitat
Azorella pedunculata
This cushion plant exemplifies one of the páramo's most distinctive growth forms. Its densely packed rosettes fuse into domelike mats that provide a buffer from wind, cold, and grazing. Beneath its tough outer surface lies a dense core of old leaves and stems, creating an insulating microhabitat that conserves heat and moisture, sustaining both the plant and tiny organisms within.

16 / Spurring diversification
Halenia weddelliana
In the páramo, this dwarf shrub brightens the landscape with striking yellow flowers tipped by nectar spurs. Each spur stores nectar at its base, accessible only to long-tongued pollinators. This tight fit between flower and visitor ensures efficient pollination and likely fuels the diversification of this species across the high Andes, where specialized pollinator relationships drive evolutionary innovation.

17 / Ancient roots
Huperzia saururus
This firmoss is related to tropical ephiphytes with a 425-million-year-old lineage. However, it has reinvented itself to thrive in open ground at high elevations. Instead of its ancestors' long stems and trailing habits, designed for humid environments, it forms low mats of erect stems with spirals of tiny fleshy leaves that have UV-protective pigments. These adaptations protect it against intense sun, wind, and water loss.

11 /

Star-shaped flowers have six tepals

· PÁRAMO ·

12 /
13 /
14 /
15 /
16 /
17 /

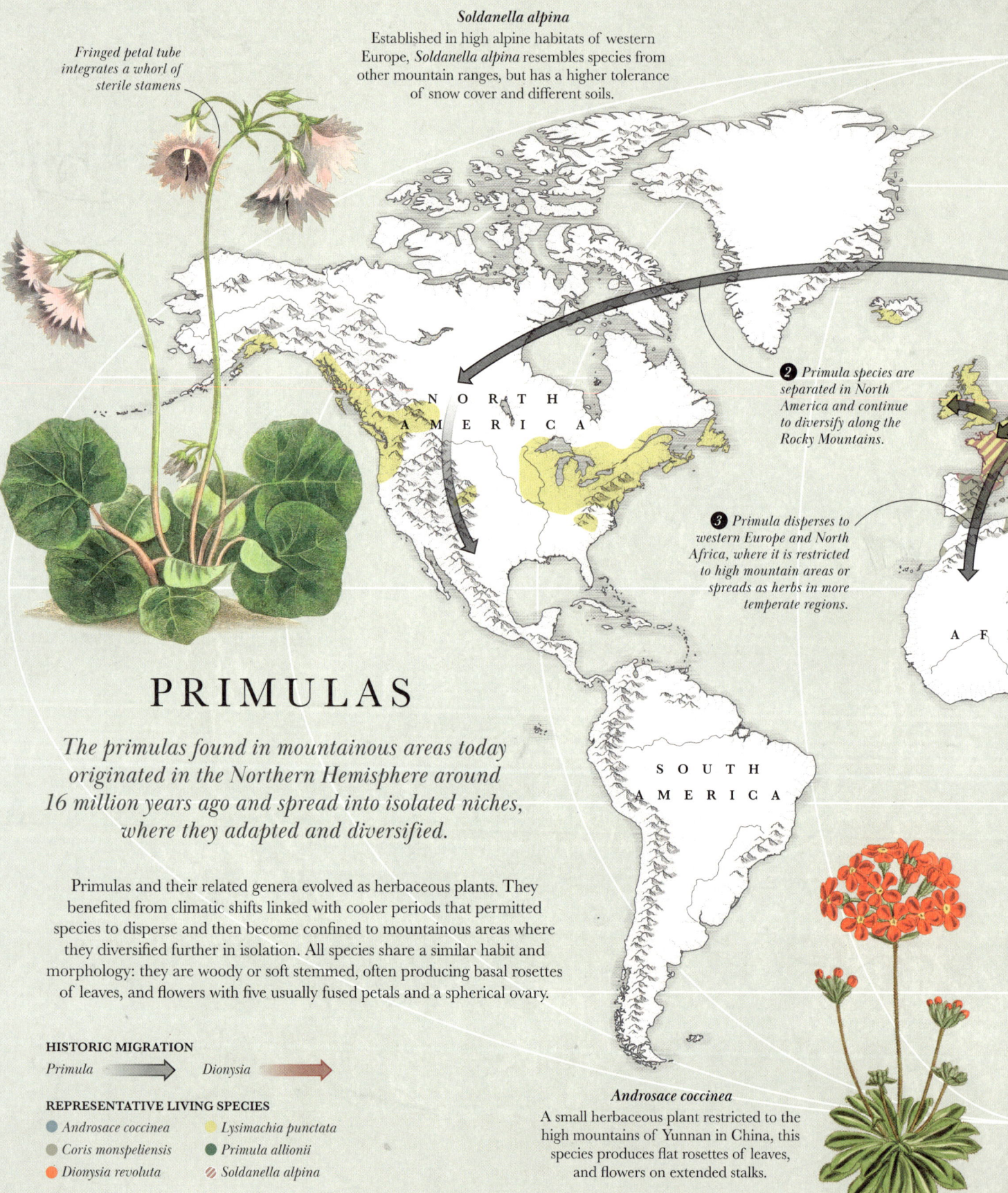

Soldanella alpina
Established in high alpine habitats of western Europe, *Soldanella alpina* resembles species from other mountain ranges, but has a higher tolerance of snow cover and different soils.

Fringed petal tube integrates a whorl of sterile stamens

2 Primula species are separated in North America and continue to diversify along the Rocky Mountains.

3 Primula disperses to western Europe and North Africa, where it is restricted to high mountain areas or spreads as herbs in more temperate regions.

PRIMULAS

The primulas found in mountainous areas today originated in the Northern Hemisphere around 16 million years ago and spread into isolated niches, where they adapted and diversified.

Primulas and their related genera evolved as herbaceous plants. They benefited from climatic shifts linked with cooler periods that permitted species to disperse and then become confined to mountainous areas where they diversified further in isolation. All species share a similar habit and morphology: they are woody or soft stemmed, often producing basal rosettes of leaves, and flowers with five usually fused petals and a spherical ovary.

HISTORIC MIGRATION
Primula   Dionysia

REPRESENTATIVE LIVING SPECIES
- *Androsace coccinea*
- *Coris monspeliensis*
- *Dionysia revoluta*
- *Lysimachia punctata*
- *Primula allionii*
- *Soldanella alpina*

Androsace coccinea
A small herbaceous plant restricted to the high mountains of Yunnan in China, this species produces flat rosettes of leaves, and flowers on extended stalks.

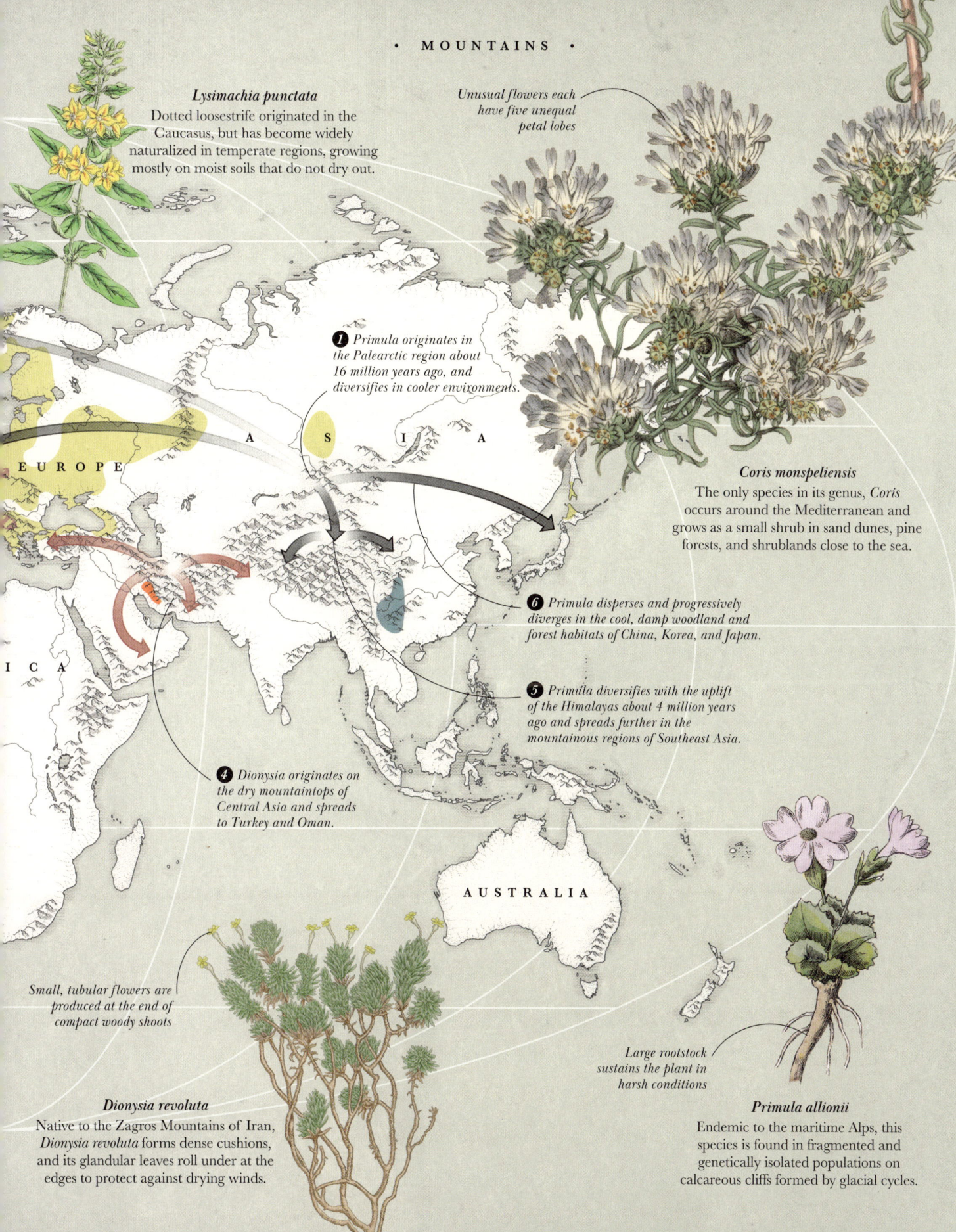

Chapter 5
WETLANDS

Fickle tides

Tidal range is the difference in height between the high water and low water levels of a tidal cycle. These ranges create different zones with different challenges for plants. Plant life must adapt to varying levels of salinity, water depth, tidal inundation, and soil conditions.

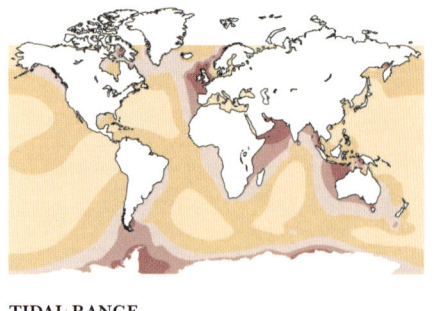

TIDAL RANGE
- 0–2 mm
- 2–4 mm
- 4–6 mm
- 6–8 mm
- 8–10 mm

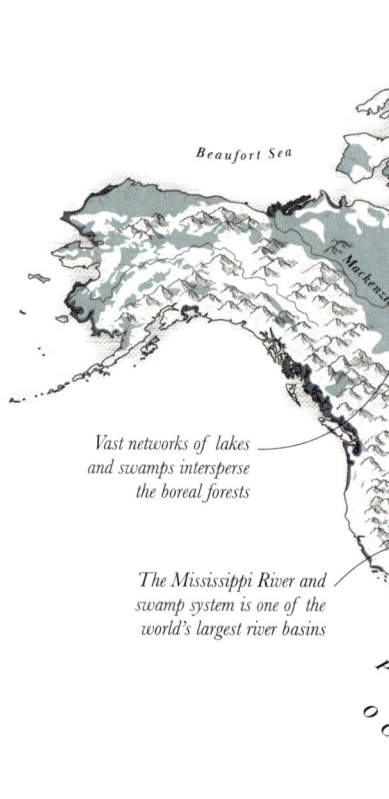

Vast networks of lakes and swamps intersperse the boreal forests

The Mississippi River and swamp system is one of the world's largest river basins

The Pantanal is a seasonally flooded mosaic of wetland, teeming with biodiversity

WETLANDS

Waterbodies found in freshwater and marine biomes provide support to plants, but light is only present in their upper reaches. Water flows – with tides, gravity, or currents, and in marine environments, storms and salt – add to the challenges of survival.

SAPPHIRE PLANET

Oceans cover around 70 per cent of the Earth's surface, while freshwater areas cover less than 2 per cent. Both are teeming with photosynthetic life – whether free floating, anchored to rocks, or in sand and mud. In many areas, freshwater wetlands intermingle with grasslands or forests, forming marshes and swamplands – especially in the tropics and taiga.

KEY
- Coastal wetlands
- Freshwater wetlands
- Average temperature

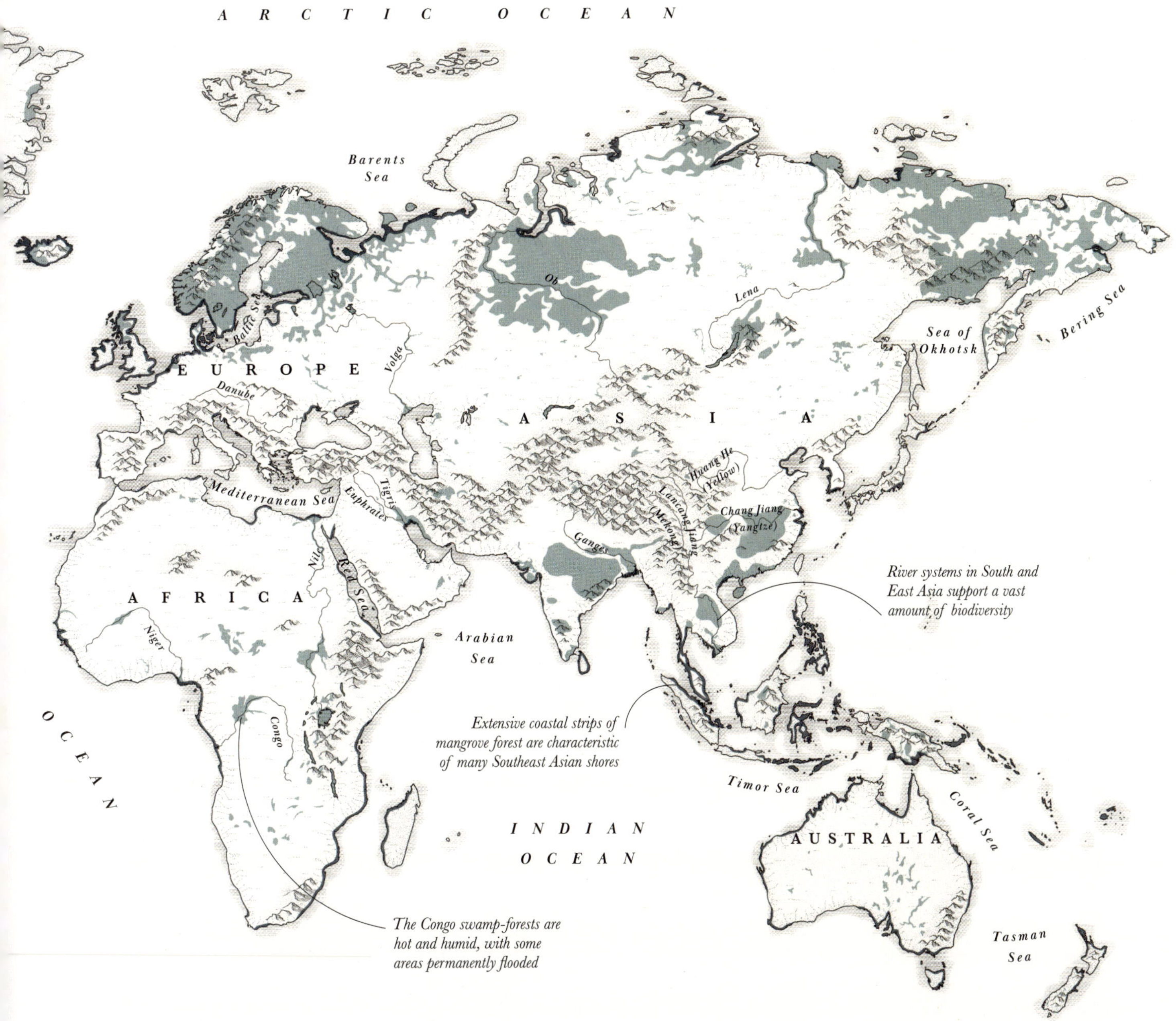

ARCTIC OCEAN
Barents Sea
Baltic Sea
EUROPE
Danube
Volga
Ob
Lena
Sea of Okhotsk
Bering Sea
ASIA
Mediterranean Sea
Euphrates
Tigris
Huang He (Yellow)
Lancang Jiang (Mekong)
Chang Jiang (Yangtze)
Ganges
Nile
Red Sea
AFRICA
Niger
Congo
Arabian Sea
River systems in South and East Asia support a vast amount of biodiversity
Extensive coastal strips of mangrove forest are characteristic of many Southeast Asian shores
OCEAN
Timor Sea
Coral Sea
INDIAN OCEAN
AUSTRALIA
The Congo swamp-forests are hot and humid, with some areas permanently flooded
Tasman Sea

Oscillating temperatures

Large-scale marine influences, such as currents and the cycles of El Niño and La Niña, can change ocean temperatures from year to year, and often influence the climate on land. Temperatures in freshwater environments may vary radically through the seasons.

1995
2000
2005
2010
2015
2020
Temperature (°C)
20
15
10
5
0
-10
COASTAL WETLANDS

J F M A M J J A S O N D
Temperature (°C)
20
15
10
5
0
-5
-10
FRESHWATER WETLANDS

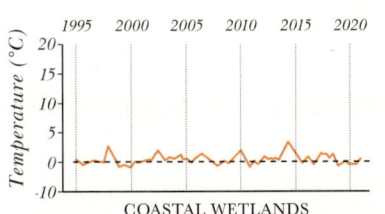

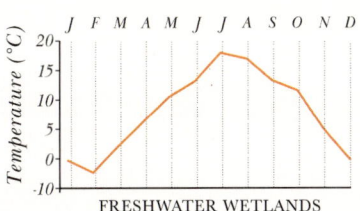

ADAPTATIONS

AIR AND MOVEMENT

Freshwater wetlands encompass bogs and marshes, river banks and lake margins as well as open water. Plants in these environments may escape threats from disease, competition, and herbivores found on dry land, but they must contend with different challenges – including how to reach sunlight or stay upright in fast-flowing conditions; how to access nutrients in waterlogged soils; and how to reproduce in the watery environment. Common to all is the issue presented by a lack of oxygen, which is essential for cellular respiration. The solutions are many and varied, as plants have evolved a range of specialist strategies to thrive at different depths.

fig. 1 Myrica gale; fig. 2 Iris pseudacorus; fig. 3 Isoetes lacustris; fig. 4 Nymphaea alba; fig. 5 Utricularia minor; fig. 6 Typha latifolia; fig. 7 Salvinia natans; fig. 8 Nuphar lutea; fig. 9 Ranunculus aquatilis

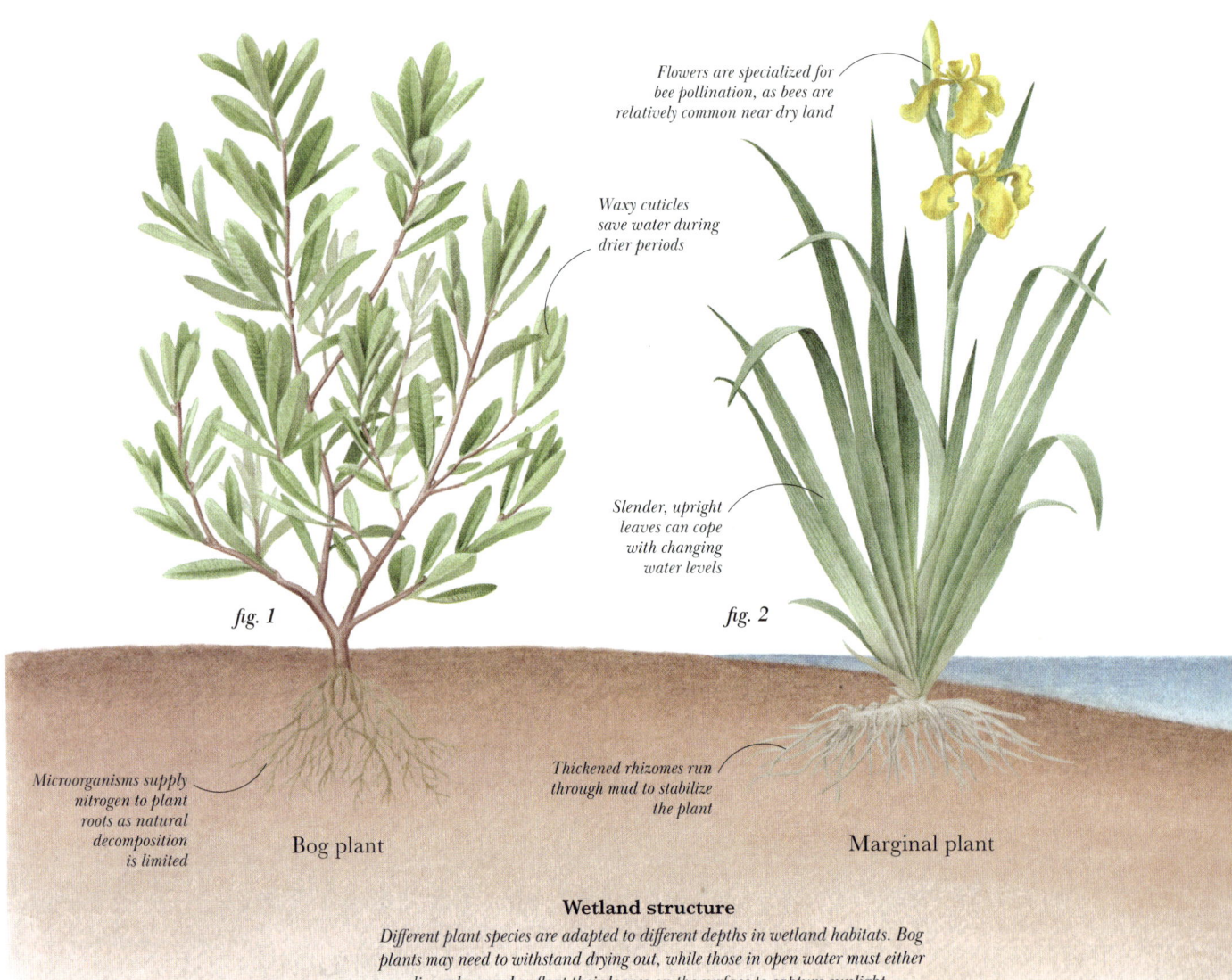

Wetland structure

Different plant species are adapted to different depths in wetland habitats. Bog plants may need to withstand drying out, while those in open water must either live submerged or float their leaves on the surface to capture sunlight.

WETLANDS

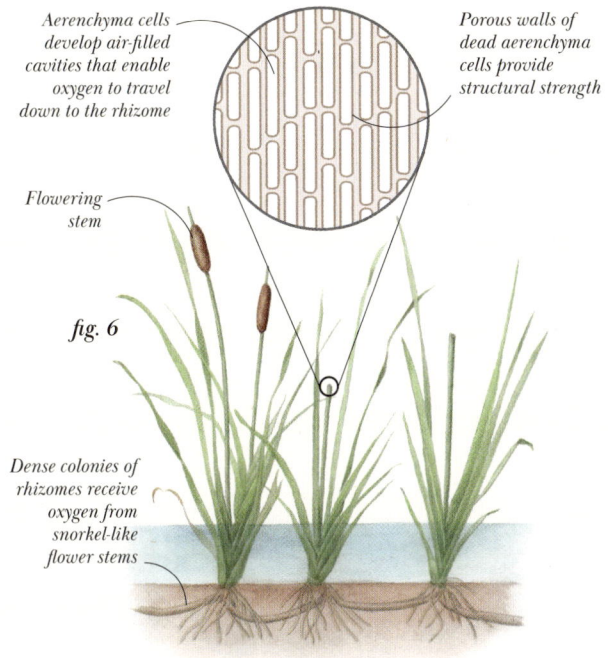

Bringing air to roots

Space for air

Plants release oxygen through photosynthesis in green leaves and shoots, but little of this oxygen is made available to non-photosynthetic roots. In oxygen-poor soils, this can be a big challenge, because oxygen is essential to break down sugars for healthy growth. Plants have specialized cells called aerenchyma to facilitate this.

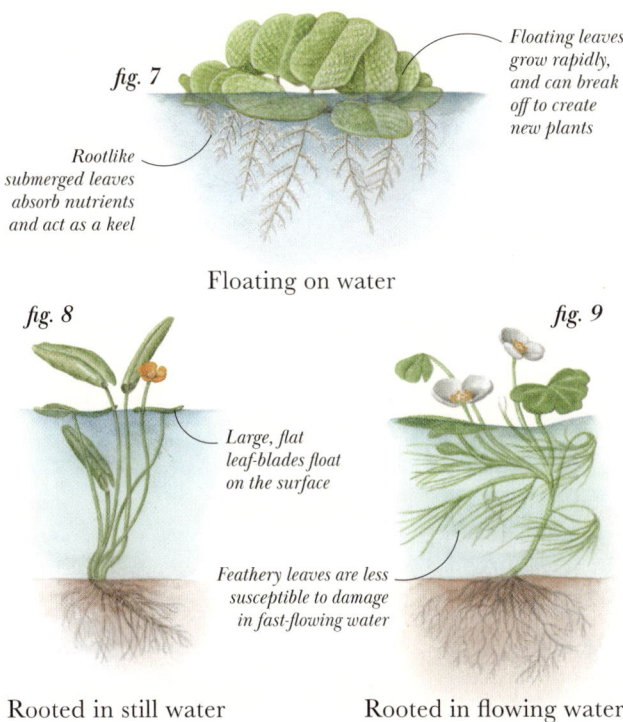

Floating on water

Rooted in still water   Rooted in flowing water

Life on the water

Unrooted water plants grow and reproduce in areas of still water, and are dispersed downstream if they drift into flowing water. Plants with strong anchoring roots are found in both still and flowing water; some have adapted to cope with the ebb and flow of fast-moving currents.

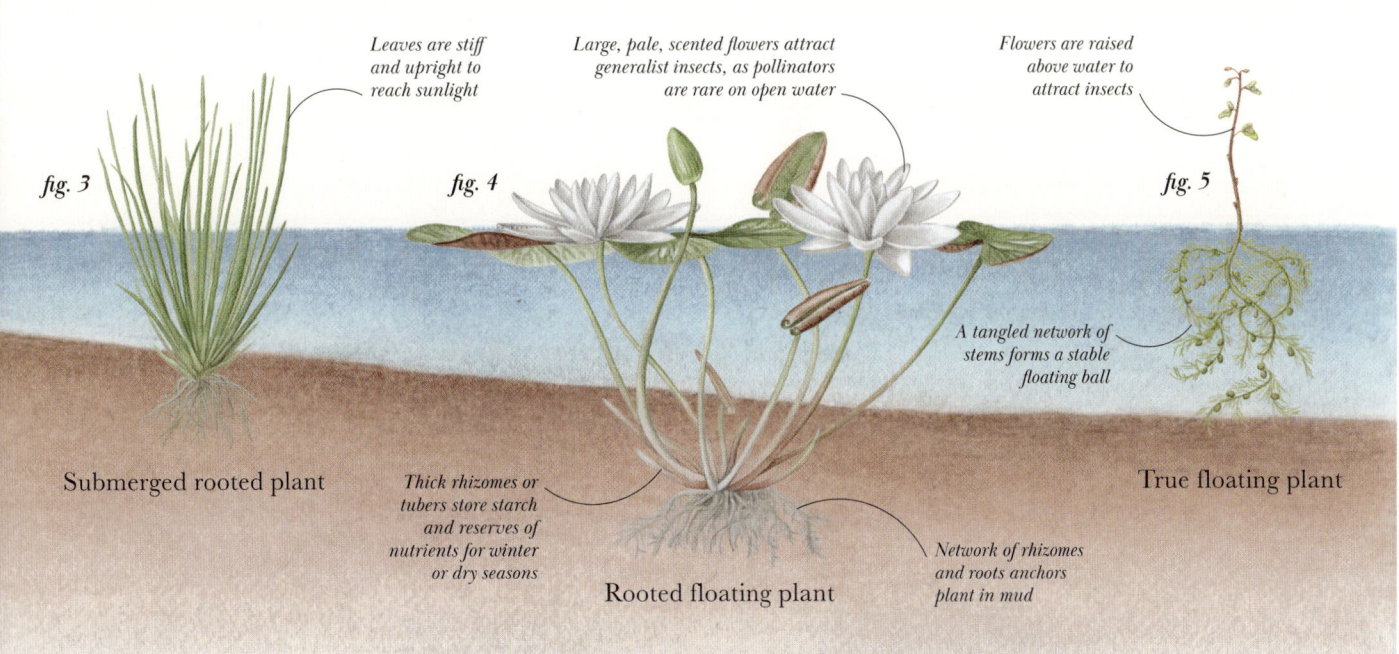

Submerged rooted plant   Rooted floating plant   True floating plant

ADAPTATIONS
SALT AND BUOYANCY

Marine environments are among the most challenging habitats for flowering plants. Not only does salt draw precious water out of their cells – interfering with many cellular processes – but to access sunlight, plants must remain upright and keep a stable footing in shifting water and substrates. A wealth of adaptations, both striking and subtle, allow plants to cope with the liminal lifestyle on offer in coastal habitats. Marine algae such as seaweeds that remain in the water face similar challenges – and those plants that live perilously close to the tideline must also contend with the prospect of drying out at low tides.

fig. 1 Acanthus ilicifolius; *fig. 2* Salicornia europaea; *fig. 3* Bruguiera cylindrica; *fig. 4* Cocos nucifera; *fig. 5* Entada gigas; *fig. 6* Fucus sp.; *fig. 7* Rhizophora mangle

Surviving salt

Salt draws water out of cells through osmosis, which can cause them to collapse. Many coastal and marine specialists cope by transporting salt to areas where it can be excreted or stored separately, or prevent salt from entering their tissues at all.

Salt crystals are excreted from sap onto leaf surfaces

Salt collects in stem segments that are shed as they become too salty

fig. 2

Storing salt

A waxy strip prevents salt entering the xylem and phloem

fig. 1

Excreting salt

Specialized cell membranes prevent salt entering cells

fig. 3

Excluding salt

WETLANDS

Woody seed-coat prevents salt water from entering the seed

Internal air chamber allows fruit to float

fig. 4

Floating fruit

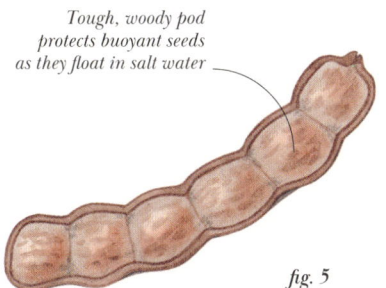

Tough, woody pod protects buoyant seeds as they float in salt water

fig. 5

Drifting seed

Life adrift

Seeds and fruits that float are common in many coastal plants. With long dormancy periods and tough, mostly watertight coatings, it is the ideal way to for plants to colonize new areas. Some seeds drift thousands of kilometres, resulting in species with extremely wide distributions.

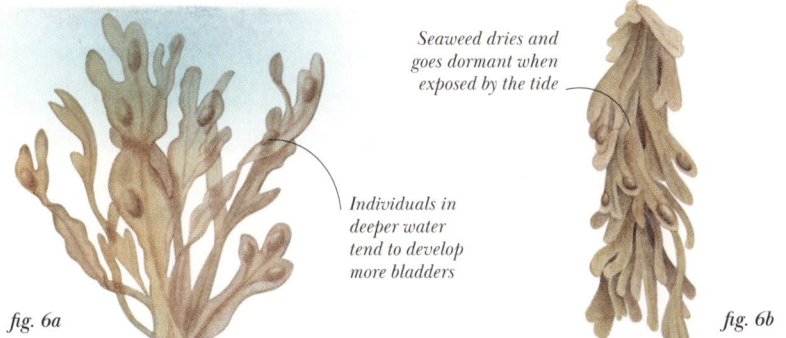

Seaweed dries and goes dormant when exposed by the tide

Individuals in deeper water tend to develop more bladders

fig. 6a

Submersion

fig. 6b

Exposure

Shifting tides

Tides, currents, and storms encourage plants to find a stable base in order to stay upright. However, some algae take a different approach, with "buoyancy aids" that raise them up when they are submerged in high tides, and specialized physiologies to cope when they are exposed by low tides.

fig. 7

Downward-growing branches called rhizophores act like prop roots to stabilize the whole plant in shifting substrates and in the ebb and flow of tides

Rhizophores form true, branching roots when they hit the silt or sand beneath the tree

Rooting down

PLANTS OF
FRESHWATER WETLANDS

The freshwater biome covers a tiny fraction of Earth's surface but harbours an abundance of life. From the seasonally frozen lakes of the far north to temperate swamplands and mighty rivers threading through tropical lowlands, freshwater habitats present plants with a range of challenges. Factors such as the clarity and flow of water, pH, oxygen, availability of carbon dioxide, and the degree of decomposition in swamplands all have a bearing on the adaptations and diversity of wetland plants.

Bright golden yellow stamens have blue tips to attract bees

1 / Blue beauty
Nymphaea nouchali var. *caerulea*
The sacred blue waterlily inhabits still or slow-flowing water throughout southern and eastern Africa. Despite the common name, the large, showy flowers are not always blue, but sometimes feature white, mauve, or pinkish-blue petals. Although the blooms can self-fertilize, their sweet scent also attracts bees, beetles, and other pollinators.

Leaf edges are slightly curved inwards, to help keep them afloat

FRESHWATER WETLANDS

2/ Feathery floater
Ceratophyllum demersum
This vigorous spreading aquatic plant grows almost entirely underwater. The feathery leaves surround slender, brittle stems that readily break off and float away to colonize new areas. The plant rarely anchors itself to the substrate and tends to enjoy a freer, more buoyant existence. The anthers detach from the male flowers and glide through the water releasing their pollen, which is received by the tiny female flowers.

3/ Flowing fern
Bolbitis heudelotii
A classic fern of stream margins – and in most cases, streams as well – the African water fern survives happily underwater, its dark fronds capturing what little sunlight filters through. A strong, horizontally growing rhizome and grasping roots allow it to cling tenaciously to rocks and logs in fast-moving streams.

4/ Far and wide
Typha latifolia
Reedmace, or cattail, is one of the most widespread aquatic plant species, and is found throughout temperate regions in the Northern Hemisphere, as well as in South America and Australia. Each spike produces up to a quarter of a million hairy seeds designed to catch the wind – which allows them to travel huge distances – or latch onto the feathers of migratory birds.

5/ Safety feature
Pinguicula vulgaris
This delicate little carnivore catches small insects such as flies on the sticky surface of its starlike rosette of leaves. Although its pollinators are typically bees that are too large to be caught by the leaves, the purple flower is held high above the plant to avoid any unfortunate mishaps.

6/ Rapid expansion
Salvinia natans
The floating fern looks nothing like a typical fern. Individual stemlike rhizomes sit on the water's surface, producing pairs of floating leaves that are filled with air pockets and covered with protective hairs. These rhizomes branch, creating a fast-growing netlike structure that can sometimes completely cover ponds and lakes. Pieces that break off can give rise to clonal offspring. Below the water's surface is a different rootlike leaf that provides a stabilizing keel and absorbs minerals from the water.

2 /   4 /

3 /   5 /

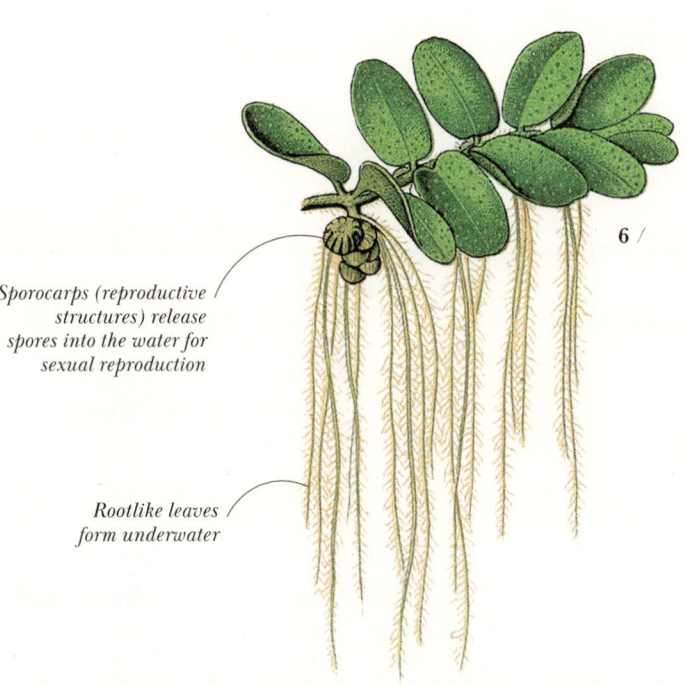

6 /

Sporocarps (reproductive structures) release spores into the water for sexual reproduction

Rootlike leaves form underwater

7 /

8 /

9 /

10 /

11 /

12 /

• FRESHWATER WETLANDS •

7/ Bayou icon
Taxodium distichum
The swamp cypress is a very adaptable deciduous conifer. It lives in fresh and brackish waters, on permanently inundated or near-dry land in the southeastern US and along the Gulf Coast. Spreading buttresses at the base give stability, while snorkel-like pneumatophores most likely bring oxygen to the roots in waterlogged soil.

8/ Heads held high
Nelumbo nucifera
Although superficially similar to a waterlily, the sacred lotus, native to Asia, is only distantly related. Unlike the floating leaves and blooms of true waterlilies, the umbrellalike leaves and elegant flowers are held high above the water. Lotus leaves are cleverly adapted to a muddy environment: their superhydrophobic properties repel water, which rolls off the surface in spherical droplets that clean the leaves and enable them to function effectively.

9/ Absorption specialist
Sphagnum capillifolium
Sphagna can cope with some desiccation but are not as tolerant as many other mosses. To help keep them moist, specialized hyaline cells allow species of sphagnum to retain up to 25 times their dry weight in water, ensuring that the peatland habitats in which they are found remain wet for longer during dry periods. The water inhibits decomposition, and leads to the buildup of huge layers of peat over time.

10/ Tiny traveller
Lemna minor
This extremely widespread tiny duckweed comprises just a few leaves and an occasional minuscule flower, with a short root to keep it stable in its stillwater habitats. The plants adhere to the legs or get lodged between the feathers of migrating waterbirds, which helps to disperse this duckweed over vast distances.

11/ Invisible lure
Sarracenia purpurea
This pitcher plant glows brightly with reflected ultraviolet light. Invisible to human eyes, the light is irresistible to insects – drawing them to the lip of the pitcher trap, where a slippery wax surface causes them to lose their footing and drop into the digestive enzymes below. Downward facing hairs in the throat of the trap prevent escape, leaving the prey to a grisly fate.

12/ Multi-functional flap
Iris ensata
This exquisite iris from East Asia forms dense mats of thick rhizomes that give rise to tufts of slender leaves. The female stigma sits behind a small flap; when a bee visits the flower, the flap scrapes pollen brought from other flowers off the bee's back, pollinating the flower. The bee then brushes past the male anther, picking up new pollen, and as the bee withdraws, the flap covers the stigma, preventing its pollen from landing on its stigma. All of this encourages cross-pollination, increasing genetic diversity and building resilience.

13/ Electrically charged
Dionaea muscipula
This charismatic little carnivore, commonly known as the Venus flytrap, is native only to the subtropical wetlands of North and South Carolina in the eastern US. When a fly – attracted to the bright red leaf-lobes – touches one or more of the six trigger hairs, an electrical charge builds up in the leaf. After multiple "hits", the charge is strong enough to trap and enclose the victim at lightning speed. The long, fingerlike cilia act as bars on a cage that gradually interlock as the insect struggles.

13/

Stiff, long cilia appear like spiny teeth

Trap reopens around 10 days after ensnaring prey

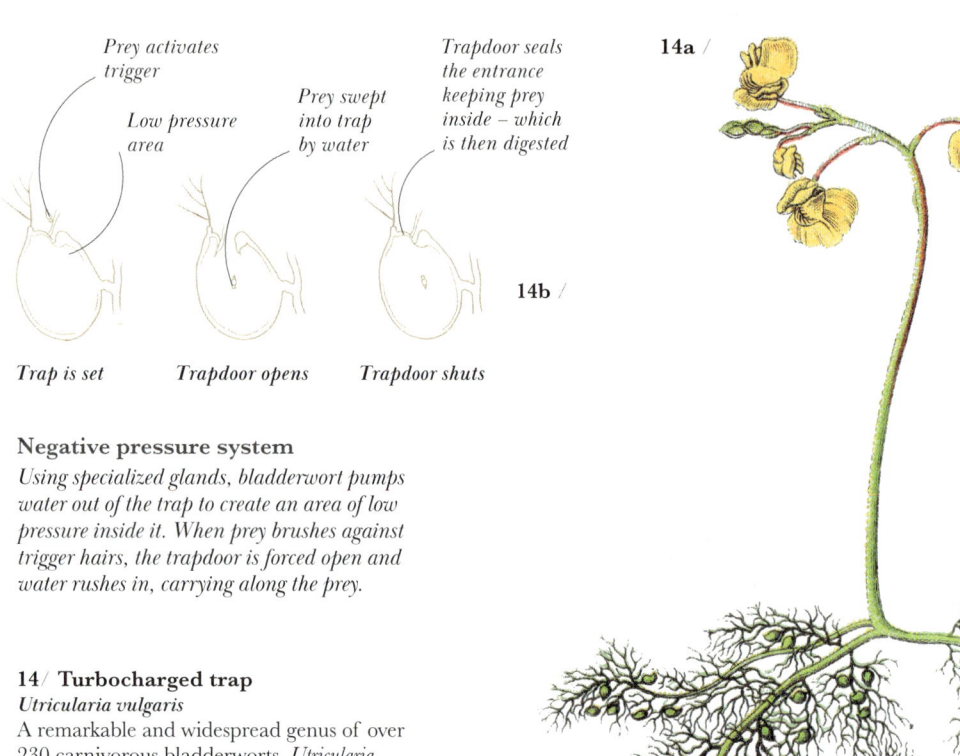

Prey activates trigger
Low pressure area
Prey swept into trap by water
Trapdoor seals the entrance keeping prey inside – which is then digested

14a
14b

Trap is set | Trapdoor opens | Trapdoor shuts

Bladder traps attach to branching stems underwater

Negative pressure system
Using specialized glands, bladderwort pumps water out of the trap to create an area of low pressure inside it. When prey brushes against trigger hairs, the trapdoor is forced open and water rushes in, carrying along the prey.

14/ Turbocharged trap
Utricularia vulgaris
A remarkable and widespread genus of over 230 carnivorous bladderworts, *Utricularia* spp. can be free-floating, tangled in other aquatic vegetation, or may have a terrestrial lifestyle. They appear to wait, poised for water fleas, amoebae, and other tiny water creatures to swim past. Then, in less than a millisecond, they open and close their traps using a vacuumlike mechanism to suck in the unsuspecting prey.

15/ Life in the fast lane
Ranunculus aquatilis
Native in most of Europe, western North America, and North Africa, common water-crowfoot produces two types of leaves to cope with life in flowing streams. The submerged leaves are finely dissected into almost feathery fronds that sit in the flow, while emergent leaves have more flattened blades, well-suited to capture sunlight.

16/ Buoyancy aid
Trapa natans
The water caltrop is found in still and slow-moving freshwater in warm temperate parts of Eurasia and Africa. It has spongy, air-filled bladders that keep it afloat and the leaves form a dense rosette on the surface of the water. The water caltrop may be free-floating or anchored in mud via a submerged stem. The single-seeded fruit is relatively heavy and either sinks to the bottom of still waters, or is transported short distances on slow currents.

17/ Amazon queen
Victoria amazonica
This giant of pools and still water in the Amazon watershed is unmistakeable. The vast circular floating leaves are a structural marvel, with hollow, radiating ribs supported by cross beams that combine to provide remarkable strength. Each leaf can grow up to 3 m (10 ft) wide and may even support the weight of a small child. The ribs, the stalks that join the leaves to the stems (petioles), and flower buds feature robust spines to deter herbivores.

18/ Sturdy sedge
Cyperus papyrus
The specialized aerenchyma cells running up the triangular stem of papyrus make it light yet strong, and allow this riverside sedge to keep its ruff of leaves and flowers high above the water level. Native to swamps and lake margins in Africa, papyrus – also called Nile grass – was famously used by the Ancient Egyptians to make the precursor to paper, which was produced by beating the stem's soft aerenchyma tissues flat, and then weaving it together.

• FRESHWATER WETLANDS •

15 /

16 /

18 /

17 /

"Symbols of the untamed chaos that surrounded and perpetually threatened the Egyptian world."

JANICE KAMRIN, "PAPYRUS IN ANCIENT EGYPT", 2000

Flower stems are sturdy and triangular

Spikelets contain numerous tiny, wind-pollinated flowers

Slender, arching rays carry clusters of spikelets

Papyrus
Cyperus papyrus

Widespread around rivers in central Africa, and once prevalent in the Nile Delta, the grasslike papyrus sedge (*Cyperus papyrus*) was an important resource for the Ancient Egyptians. They used the stems in the construction of reed boats and for household goods such as mats, rope, and sandals, while the woody root could be used to make bowls. Their most significant innovation, however, was the extraction of the central pith to produce a paperlike material. As a writing material, papyrus was essential for the development of written language and record-keeping. A well known example is the Ebers papyrus – a compendium of herbal knowledge and one of the oldest known records of ancient medical knowledge.

Rising each year from the annually flooded marshes, papyrus also held important symbolic meanings for Ancient Egyptians that connected with their ideas about youth and rebirth. Reminiscent of their creation story – which saw the world emerge from a limitless expanse of water – papyrus represented the limits of their own ordered world and a chaotic world beyond. The plant's ubiquitous presence around the Nile Delta also made papyrus a symbol of Lower Egypt that was strongly associated with the political power of the Pharaoh.

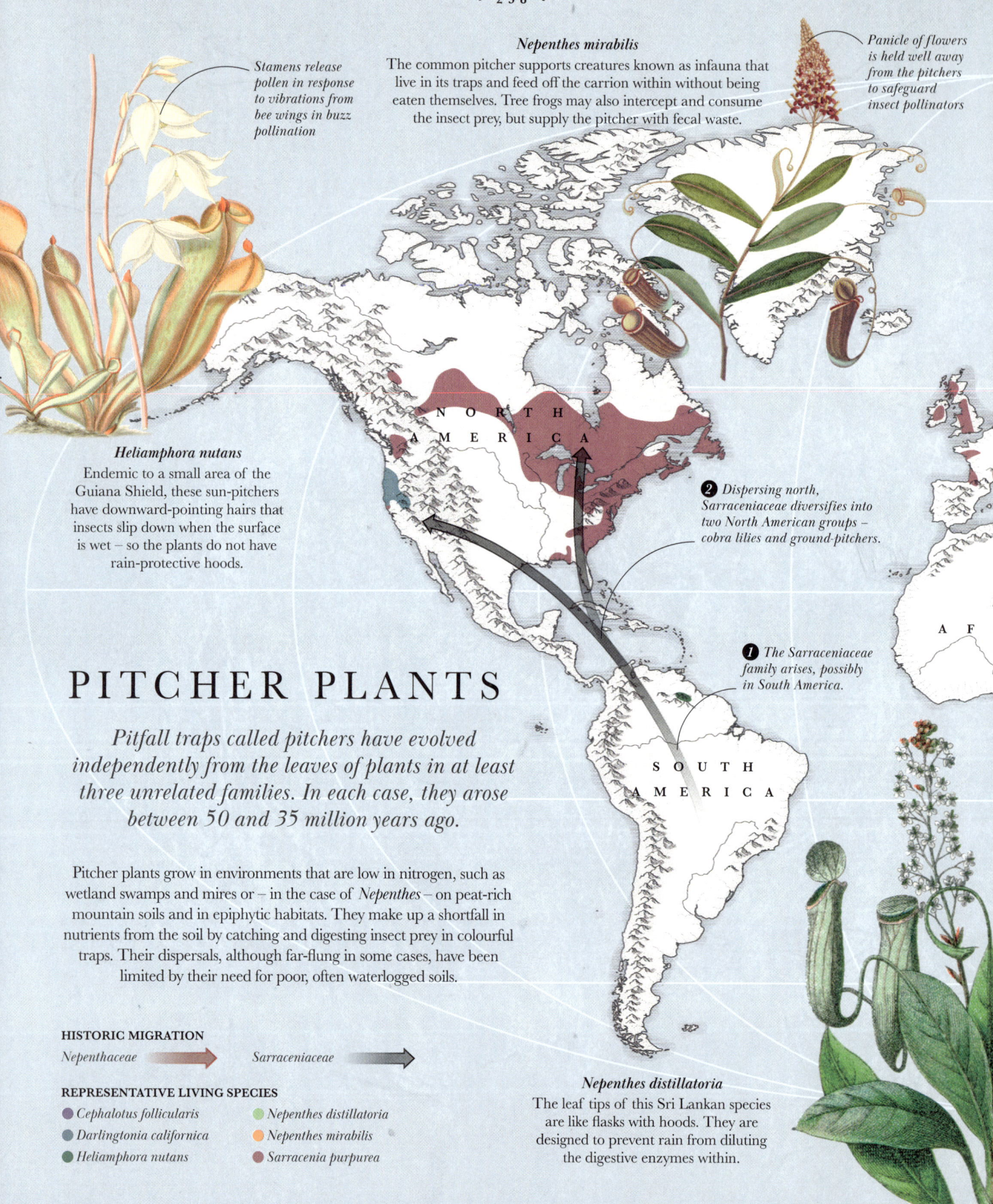

PITCHER PLANTS

Pitfall traps called pitchers have evolved independently from the leaves of plants in at least three unrelated families. In each case, they arose between 50 and 35 million years ago.

Pitcher plants grow in environments that are low in nitrogen, such as wetland swamps and mires or – in the case of *Nepenthes* – on peat-rich mountain soils and in epiphytic habitats. They make up a shortfall in nutrients from the soil by catching and digesting insect prey in colourful traps. Their dispersals, although far-flung in some cases, have been limited by their need for poor, often waterlogged soils.

HISTORIC MIGRATION
Nepenthaceae → Sarraceniaceae →

REPRESENTATIVE LIVING SPECIES
- *Cephalotus follicularis*
- *Darlingtonia californica*
- *Heliamphora nutans*
- *Nepenthes distillatoria*
- *Nepenthes mirabilis*
- *Sarracenia purpurea*

Stamens release pollen in response to vibrations from bee wings in buzz pollination

Nepenthes mirabilis
The common pitcher supports creatures known as infauna that live in its traps and feed off the carrion within without being eaten themselves. Tree frogs may also intercept and consume the insect prey, but supply the pitcher with fecal waste.

Panicle of flowers is held well away from the pitchers to safeguard insect pollinators

Heliamphora nutans
Endemic to a small area of the Guiana Shield, these sun-pitchers have downward-pointing hairs that insects slip down when the surface is wet – so the plants do not have rain-protective hoods.

❶ The Sarraceniaceae family arises, possibly in South America.

❷ Dispersing north, Sarraceniaceae diversifies into two North American groups – cobra lilies and ground-pitchers.

Nepenthes distillatoria
The leaf tips of this Sri Lankan species are like flasks with hoods. They are designed to prevent rain from diluting the digestive enzymes within.

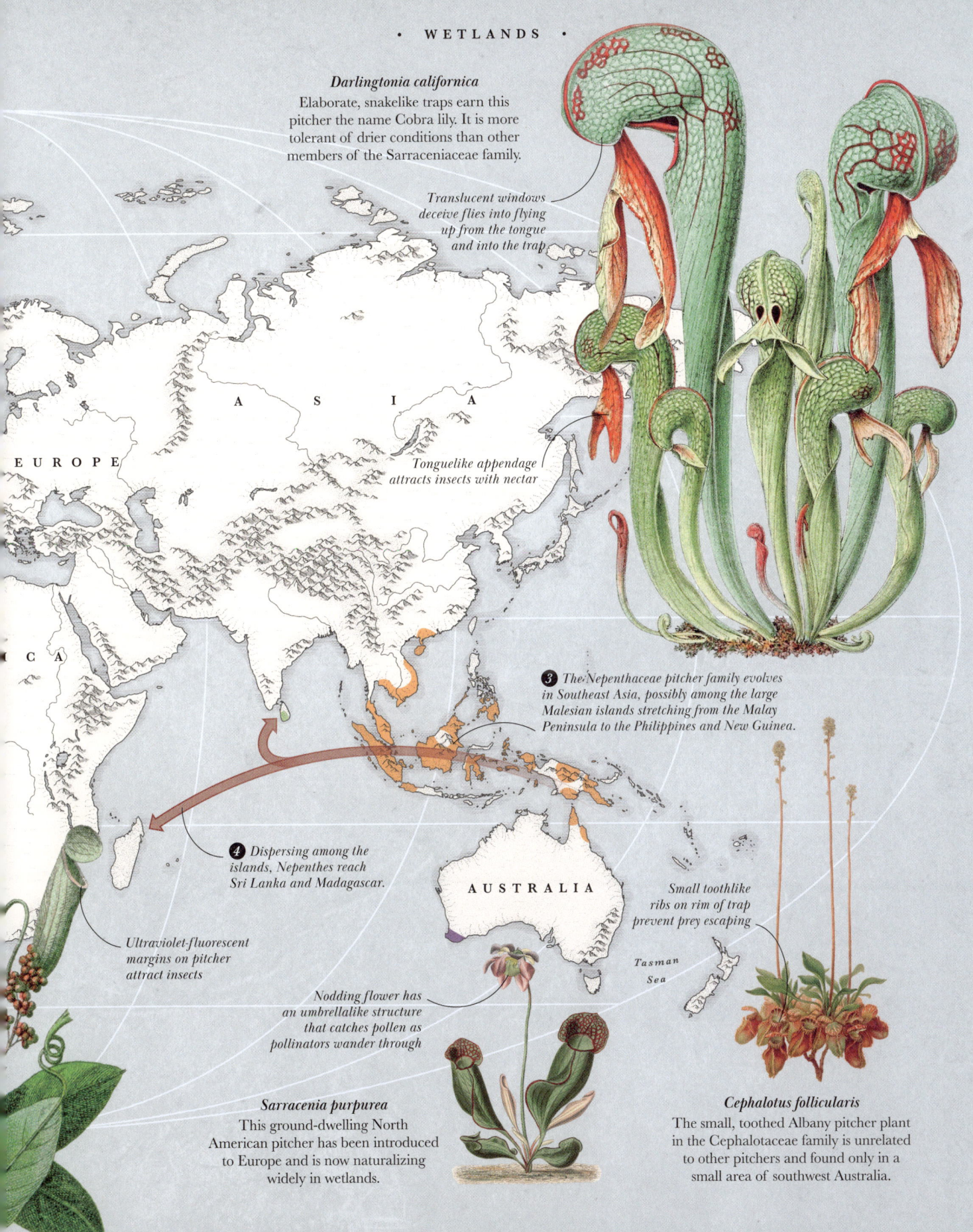

PLANTS OF
MARINE AND COASTAL WETLANDS

The marine biome encompasses more than 70 per cent of Earth's surface area, from the icy, tempestuous polar seas to the tropical doldrums. Plants and other photosynthesizers such as algae need to contend with fickle currents, tides, and storms as well as constant high salinity, to thrive here. The trade-off can be relatively little competition in such specialized environments. Marine habitats range from brackish-water estuaries, with often shifting muddy or sandy substrates, to rocky shores and cliffs, or the open ocean – all of which present distinct challenges, and opportunities, for adaptation.

Banner petal attracts pollinators

Straight pods are up to 15 cm (6 in) long and contain two to ten seeds

1/ Topsy-turvy tropical pea
Canavalia rosea
The beach bean is widely distributed throughout tropical regions thanks to its buoyant seeds that are carried huge distances on ocean currents. Its long stems, punctuated by scented pink-petalled flowers, form extensive mats on sandy and stony beaches. Unlike related peaflowers, these blooms typically hang upside-down, with the large upper petal (the banner) facing downwards and acting as a landing pad for visiting insects.

2/ Resourceful foragers
Elphidium crispum
Like other members of the Foraminifera subphylum, this single-celled organism usually acts as a microscopic predator or feeds on detritus using long strands called ectoplasmic pseudopods extruded from its cells. However, some foraminifera also host algae, sheltering them in return for energy via photosynthesis. Other foraminifera ingest algae, harvesting their photosynthetic cells in order to perform photosynthesis themselves.

3/ Fast-moving creeper
Ipomoea pes-caprae
Known as beach morning glory, this creeping vine has leaves with thick cuticles and relatively small stomata (pores) on their upper surfaces, protecting it from desiccation in the salt air. Stems can be up to 50 m (165 ft) in length, with deep roots that help to stabilize sand dunes. This ability allows other plants to make footholds in this unstable environment.

4/ Underwater forests
Laminaria hyperborea
Cuvie, as this robust brown seaweed is known, forms extensive underwater forests in the North Atlantic Ocean, supporting a wealth of marine life. Intolerant of desiccation, it lives almost completely submerged. To ensure it can reach as much sun as possible, its fronds have strong, upright stalks (stipes) that are rich in flexible algal polysaccharides. This allows the plant to sit high in the water yet remain flexible in the tides and ocean currents.

5/ Small but resilient
Pelvetia canaliculata
Channelled wrack is a hardy seaweed that lives at the very top of rocky shores, in the splash zone. It deals with being completely dry during summer neap tides (when there is the least difference between high and low tides), by slowing or shutting down some of its processes, until it is rehydrated again by larger tides, wetter weather, or storms. This mechanism is poorly understood, but it may involve accumulation of the sugar mannitol in its tissues, and the relatively high fat and protein content in this seaweed.

6/ Swimming seeds
Nypa fruticans
The nipa palm is common in tropical estuaries from Southeast Asia to Australia, but has a wide range of salt tolerance, so can be found in freshwater and neighbouring dry land as well. Its edible, globe-shaped spiky fruits contain large seeds that have a buoyant husk, making them capable drifters. The plant's large fronds, can reach up to 9-m (30-ft) long.

Bumpy, V-shaped swellings are its reproductive structures

7 /

8 /

9 /

10 /

11 /

12 /

• MARINE AND COASTAL WETLANDS •

7/ Rooting in rock
Armeria maritima
Also known as thrift or cliff rose, this clump-forming perennial is beautifully adapted to its coastal habitat – its low-growing habit and small, inrolled leaves prevent water loss in the salty drying winds. Beneath the soil, a taproot, sometimes over 1-m (3-ft) long, anchors the plant in the cracks among cliffs, allowing it to cling on in the most violent winter storms.

8/ Champion drifter
Lathyrus japonicus
The seeds of the beach pea are remarkable drifters. Early experiments by Charles Darwin involved floating a range of seeds, including those of the beach pea, in tubs of seawater, and shaking them occasionally to simulate waves. The test showed that the beach pea seed has an extraordinary ability to float, for months on end, and still remain viable for germination. This is reflected in its modern distribution – throughout the cold and temperate Northern Hemisphere, but also with possible native populations in Chile and New Zealand.

9/ Coastal conifer
Podocarpus polystachyus
Sea teak belongs to the conifer group and its particularly tough, leathery leaves make it one of the few conifers that can live in or near the sea. Its large leathery seeds have swollen, fleshy bases – the seeds and this structure are gulped down whole by birds such as pigeons, and the seeds later dispersed in their droppings.

10/ Marine meadows
Zostera marina
Common eelgrass is one of very few flowering plant species to live entirely underwater, where it forms extensive meadows with its horizontal underground stems (rhizomes). It grows on sandy seabeds at shallow depths, up to around 10 m (33 ft), because it needs to access sunlight in order to photosynthesize. Its swathes of narrow leaves provide shelter for seahorses and nurseries for small fish as well as food for birds in winter.

11/ Walking wonder
Pandanus utilis
The common screwpine is an evergreen palmlike tree with incredible examples of stilt roots that help it maintain a stable footing in shifting sands. The original root sometimes dies back, leaving the main stem propped up by a pyramid, or skirt, of roots. If the plant begins to keel over, more roots will emerge from the falling side. Over time this can lead to the plant "walking" – very slowly – down the beach.

12/ Flex with the flow
Laminaria digitata
This large brown alga, also known as tangle, lives higher up the shore than its relative *Laminara hyperborea*, so is more likely to be exposed at low tides. When not supported by water, its flat stipes (stalks) bend at the tips, so that its palm-like fronds dip into the sea. This movement ensures that most of the plant remains immersed, preventing desiccation.

13/ Glass menagerie
Diatoms
These microscopic algae are found in virtually all aquatic habitats. They can be broadly categorized into centric diatoms – displaying radial symmetry – and pennate diatoms – exhibiting one line of symmetry. All of them protect their single cell in a two-part glasslike structure made of silica, called a frustule. Some have evolved special features such as spines, or the ability to form chains, to slow sinking – allowing them to stay in sunlit surface waters to photosynthesize. Others regulate buoyancy by altering their internal composition, allowing them to sink and rise as needed.

Striae, or pores, allow nutrients in and send waste out

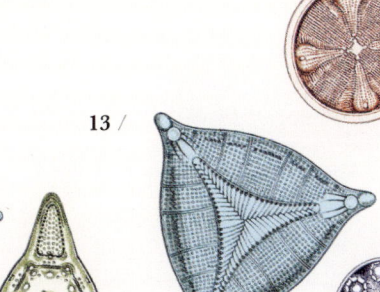

Centric diatoms are often round, triangular, or star-shaped

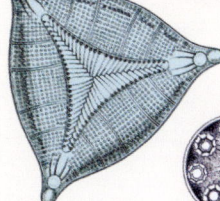

14/ Prodigious pods
Entada gigas
This liana (woody climber) in the pea family is native to Central America, northern South America, the Caribbean, and tropical Africa. It rambles through trees near coasts and the mouth of rivers. The bottle-brushlike flower heads produce a few enormous pods – some reaching over 1 m (3 ft) in length. These can fall as a whole, or split open, dropping the large, buoyant seeds into the water. The seeds can be carried huge distances across the Atlantic.

15/ Dune pioneer
Calamagrostis arenaria
Marram is a robust coastal grass of sand dune ridges found throughout the temperate Northern Hemisphere. The tightly-rolled cylindrical leaves trap humid air, preventing the drying sea air from stripping moisture from the plant. A key pioneer species, its specialized growth pattern helps to stabilize sand dunes.

16/ Isolated giant
Lodoicea maldivica
Outright winner of the largest and slowest-developing seed in the world, the massive heart-shaped seeds of coco-de mer can weigh over 30 kg (66 lbs). Their density means that, in spite of this plant's often coastal habitat, the seeds do not float, making this species endemic and restricting it to a few islands in the Seychelles. A lack of predators or disease appear to have caused it to develop the gigantism often seen in island plants.

17/ Floating giants
Macrocystis pyrifera
Giant kelp is the largest brown alga, forming extensive forests in the Eastern Pacific. Its stemlike stipes can extend over 50 m (165 ft), producing leaflike blades along their length. Towards the base of the stipe, these blades develop reproductive structures (sporophylls), whereas the ones higher up the stipe form air bladders, which keep the alga afloat.

18/ Water filter
Rhizophora mangle
Native to the tropical and subtropical coasts of North, Central, and South America, and West Africa, red mangrove has solved the challenge of living in its marine environments. It has stilt roots that grow from the trunk to keep it upright in the shifting silt and sand. These roots have a special impermeable membrane that acts as a filtration system, preventing salt from being absorbed into the plant's veins alongside the water it needs to take in for survival.

19/ Ephemeral colour
Hibiscus tiliaceus
The striking petals of this small tree, also called sea cottonwood, draw in a range of pollinators to the large, showy flowers. Each flower only usually lasts a day and, in that time, they change in colour from bright yellow, to orange, to red before dropping. Although the seeds are small, they can float for extended periods, and indeed the species is found throughout tropical coasts, including almost every Pacific island.

20/ Armoured photosynthesizers
Emiliania huxleyi
Microscopic coccolithophores are single-celled organisms, and their complex cell is surrounded by protective calcium carbonate plates, called coccoliths, forming a structure called a coccosphere. Their huge numbers in the Antarctic Southern Ocean mean the calcium carbonate can be detected from space. Its cell has a few large chloroplasts, used for photosynthesis. Unlike green plants, these chloroplasts are brown, allowing the coccolithophore to capture light at a range of depths as it migrates up and down in the water column each day.

14 /

Seeds contain a hollow cavity, which gives them buoyancy

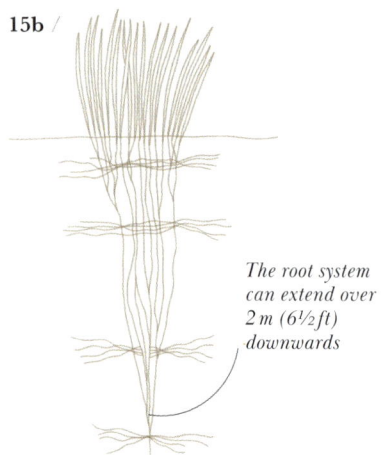

15b /

The root system can extend over 2 m (6½ ft) downwards

Sand binder
Marram anchors itself to a sand dune through an underground network of fibrous roots that extend both vertically and horizontally.

• MARINE AND COASTAL WETLANDS •

15a /
17 /
18 /
19 /
16 /
20 /

Giant waterlily
Victoria amazonica

An origin story by the Tupi-Guarani people of Brazil tells of Naiá, a girl who falls in love with the Moon who, seeing its reflection in the water, dives in to be with her heart's desire. Naiá tragically drowns and the Moon transforms her into the exquisite giant waterlily (*Victoria amazonica*). The tales associated with the plant and the names given to it – Irupé is the Guaraní name meaning a "platter on the water" – are testimony to its cultural importance throughout the Amazon basin, where it is found in still waters.

By 1847, the giant waterlily had captured the attention of plant-lovers around the world. European botanists quested through the Amazon rainforest in search of the giant, and the owners of hothouses raced to grow the plant. In the UK, Joseph Paxton brought it to flower in 1849. The spectacle of his seven-year-old daughter perched on a leaf caused a sensation, and it is thought that the leaf's structure inspired Paxton's design for the Crystal Palace in 1850. Today, these plants still elicit a sense of wonder – in the glasshouse or in the wild.

Load bearer
The vast leaves of the giant waterlily have long formed enticing platforms for those bold enough to try them, as this photograph from the 1920s shows. The leaves grow to about 3m (10ft) across, and can bear the weight of an adult human if evenly distributed.

Leaf has upturned edges with protective spines on the underside

• WETLANDS •

"Fain would I have plunged into the lake to procure specimens of the magnificent flowers and leaves."

Thomas Bridges, quoted by William Jackson Hooker, *Description of Victoria regia*, 1847

Flowers close temporarily overnight to trap beetle pollinators, releasing them the next day

Flower changes from white to pink when pollen is released

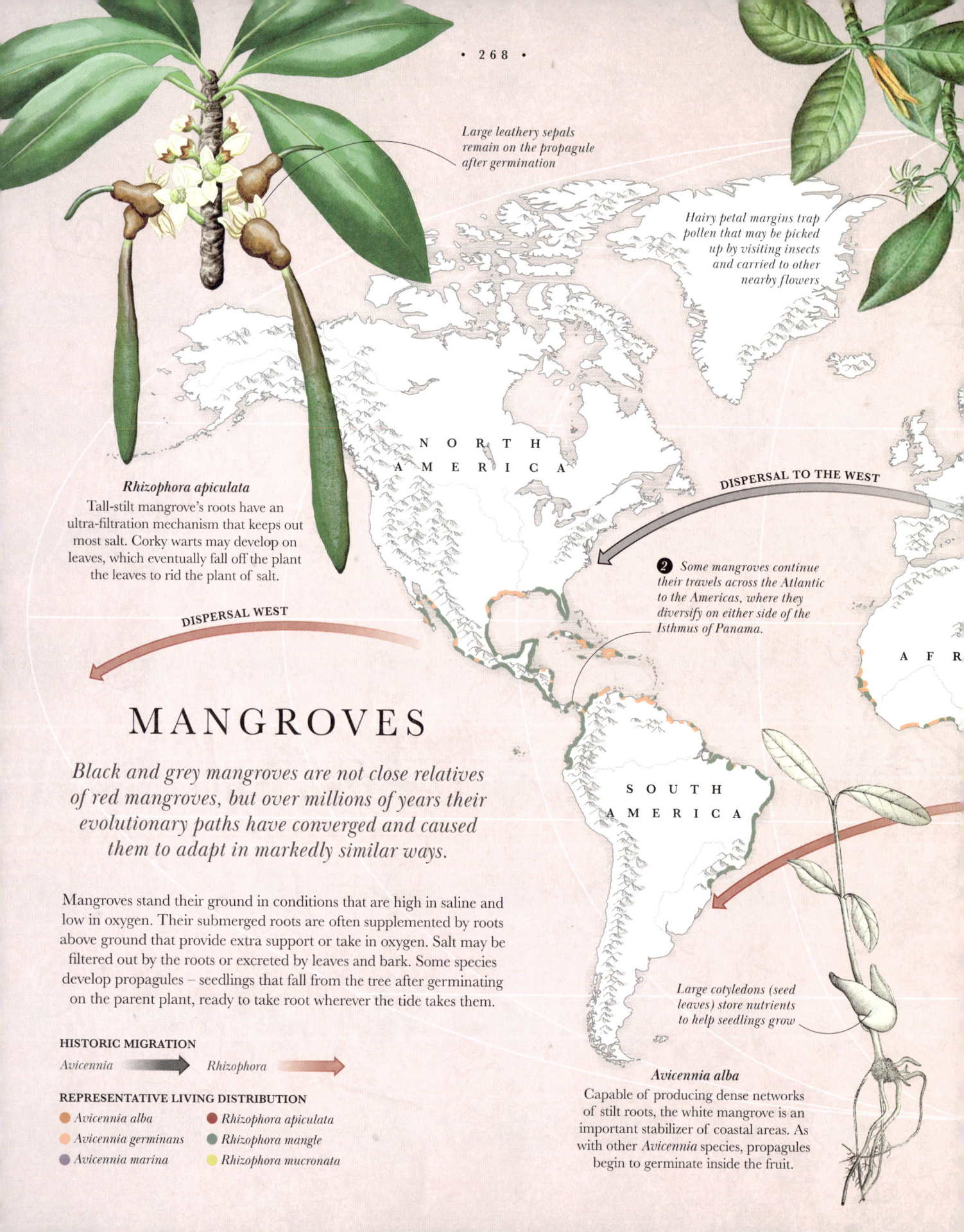

MANGROVES

Black and grey mangroves are not close relatives of red mangroves, but over millions of years their evolutionary paths have converged and caused them to adapt in markedly similar ways.

Mangroves stand their ground in conditions that are high in saline and low in oxygen. Their submerged roots are often supplemented by roots above ground that provide extra support or take in oxygen. Salt may be filtered out by the roots or excreted by leaves and bark. Some species develop propagules – seedlings that fall from the tree after germinating on the parent plant, ready to take root wherever the tide takes them.

HISTORIC MIGRATION
Avicennia ⟶   Rhizophora ⟶

REPRESENTATIVE LIVING DISTRIBUTION
- *Avicennia alba*
- *Avicennia germinans*
- *Avicennia marina*
- *Rhizophora apiculata*
- *Rhizophora mangle*
- *Rhizophora mucronata*

Rhizophora apiculata
Tall-stilt mangrove's roots have an ultra-filtration mechanism that keeps out most salt. Corky warts may develop on leaves, which eventually fall off the plant the leaves to rid the plant of salt.

Large leathery sepals remain on the propagule after germination

Hairy petal margins trap pollen that may be picked up by visiting insects and carried to other nearby flowers

DISPERSAL WEST

DISPERSAL TO THE WEST

❷ Some mangroves continue their travels across the Atlantic to the Americas, where they diversify on either side of the Isthmus of Panama.

Avicennia alba
Capable of producing dense networks of stilt roots, the white mangrove is an important stabilizer of coastal areas. As with other *Avicennia* species, propagules begin to germinate inside the fruit.

Large cotyledons (seed leaves) store nutrients to help seedlings grow

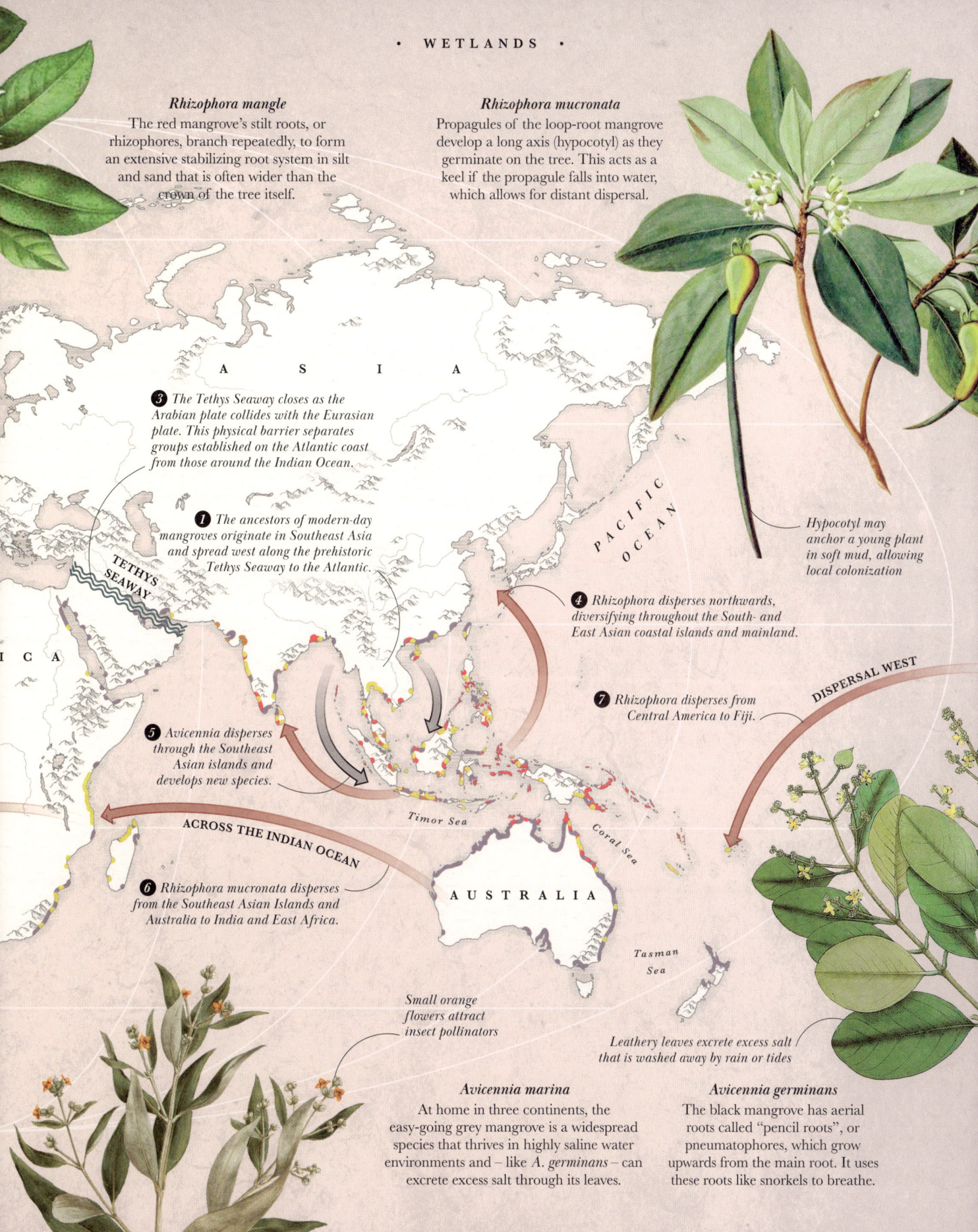

ALGAL BLOOM

Algae are important primary producers: their photosynthesis provides around half of Earth's oxygen each year. They also capture vast quantities of carbon and store it in the form of carbohydrates that fuel both marine and terrestrial food chains. However, sometimes algae can be too successful. An explosion of algal life – a sudden, or subtle shift in environmental conditions that favour one or a few species over others – can occur in practically any watery environment. When the conditions are right, algae can reproduce – or bloom – exponentially, and fill lakes or coastlines with vibrant colours.

The factors that trigger algal blooms vary according to the species of algae and the wider ecology of the area. Extended periods of warmth or sunshine can be significant, but the most common cause is a sudden inundation of nitrogen or phosphorus – most likely from overapplications of artificial fertilizer, or storms carrying overflows of sewage into lakes and rivers – disrupting the balance of nutrients in the water system.

Cyanobacteria – formerly and misleadingly known as blue-green algae because their dense growths can turn water a bluish-green colour – are bacteria that can perform photosynthesis; they are also a common cause of Harmful Algal Blooms (HABs) that produce a wide range of dangerous toxins. However, non-toxic species of algae that bloom also cause considerable problems, as they prevent light from reaching other algae and plants lower in the water. They commandeer dissolved carbon dioxide for photosynthesis – which starves aquatic plants of another precious resource. Finally, when the algae die, bacteria and other decomposers take up the oxygen in the water, and asphyxiate animal life.

Although in some places blooms have become a near annual event, experience suggests that intervention to tackle the causes of nutrient imbalances can gradually help the water – and the life it supports – to recover.

Troubled waters
The green hues of an algal bloom on Lake Amatitlán in Guatemala seem vibrant, but it is fuelled by pollutants, creating an environment in which cyanobacteria thrive. At its height, the bloom creates a "dead zone" in which fish and other aquatic organisms cannot survive.

Chapter 6

HUMAN ENVIRONMENTS

Nitrogen application

Artificial fertilizers containing varying quantities of nitrogen are used in many agricultural areas. However, if they are overapplied, nitrogen becomes a major pollutant, as it stimulates the growth of vigorous plants and algae, allowing them to outcompete less nitrogen-hungry species.

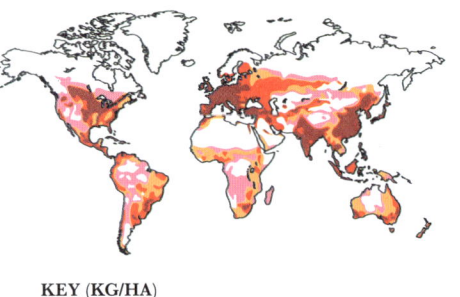

KEY (KG/HA)

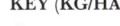

 0–50   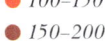 100–150
50–100   150–200

In the Great Plains, huge areas of biodiverse prairie and steppe grassland have been converted to agriculture

Areas within the Cerrado, a sprawling savanna with dry forests, have been extensively cleared for soy cultivation

HUMAN ENVIRONMENTS

Humans have had a profound impact on the environment. The creation of urban areas with globe-trotting transport links and vast tracts of agricultural land has effectively generated a new biome. For some plants, this is an existential threat; for others it is an opportunity.

ANTHROPOGENIC LANDSCAPES

Urban centres often cluster around trade hubs, while agricultural land is found in places with conditions that suit the plants that are grown as crops. Peppered throughout this human landscape, and in its more remote fringes, are networks and islands of protected areas – from small reserves to vast national parks.

Greenhouse gas emissions

Since the mid-19th century, sustaining a growing human population has caused a dramatic increase in atmospheric levels of greenhouse gases such as carbon dioxide, methane, and nitrous oxide. Their impact on global climates has significantly affected plant growth and diversity.

KEY

- Protected area
- Agricultural area
- Built-up area

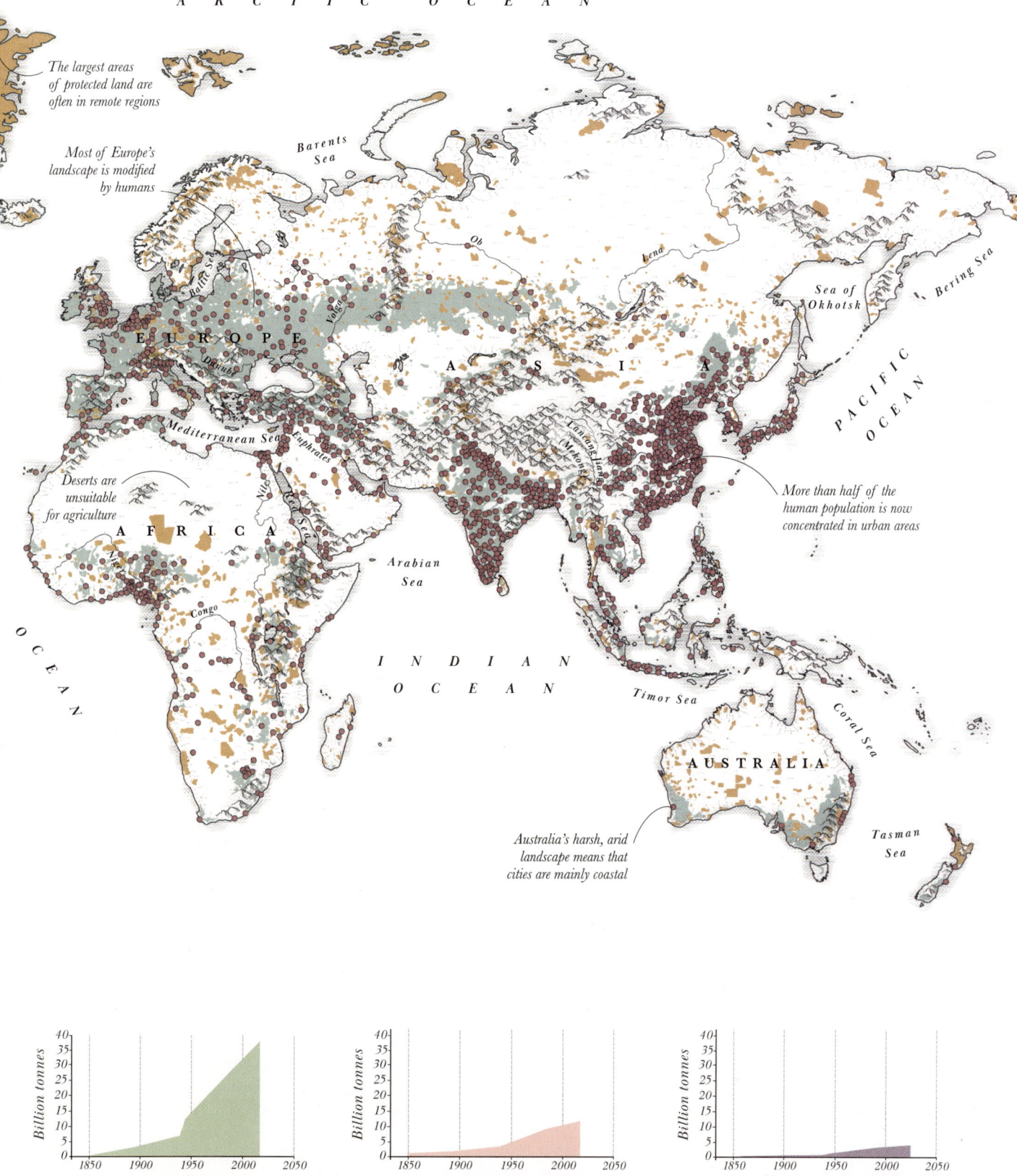

Strategies and tactics

Plants have developed suites of adaptations for different habitats that fall into three broad categories. Competitive plants are often found in areas with plenty of resources. Ruderals include fast-living annuals intent on producing seeds. Specialists adapt to particular environments, such as aquatic habitats.

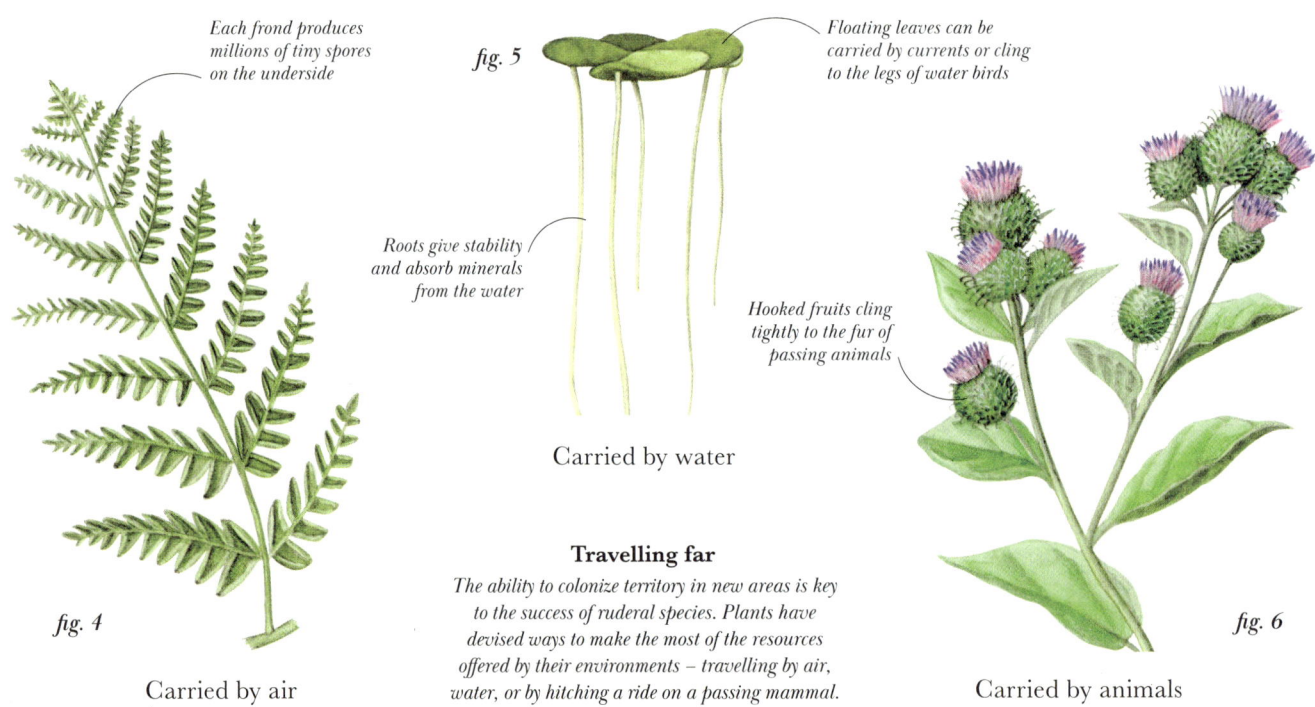

Profuse flowers produce abundant seed

Large flowers attract relatively infrequent mountain pollinators

Weedy annual plants invest little energy in roots and stems

fig. 1 — Competitive

fig. 2 — Ruderal

fig. 3 — Specialist

Each frond produces millions of tiny spores on the underside

Floating leaves can be carried by currents or cling to the legs of water birds

Roots give stability and absorb minerals from the water

Hooked fruits cling tightly to the fur of passing animals

fig. 4 — Carried by air

fig. 5 — Carried by water

fig. 6 — Carried by animals

Travelling far

The ability to colonize territory in new areas is key to the success of ruderal species. Plants have devised ways to make the most of the resources offered by their environments – travelling by air, water, or by hitching a ride on a passing mammal.

• HUMAN ENVIRONMENTS •

ADAPTATIONS
COMPETITION AND GROWTH

Plants usually have an overall strategy for life – they can be varying degrees of competitive, weedy (ruderal), or specialized for particular habitats, and often sit somewhere among these three extremes. The human biome tends to favour ruderal-competitive strategists, although croplands are geared towards providing a pampered existence for specialists. For weeds in the human environment, the key to success often hangs on the ability to disperse far and wide. Urbanization often brings such plants to new areas, where – in classic ruderal-competitive form – they go on to conquer nearby territory at a local level.

fig. 1 Paulownia tomentosa; fig. 2 Stellaria media; fig. 3 Dryas octopetala; fig. 4 Pteridium aquilinum; fig. 5 Lemna minor; fig. 6 Arctium minus; fig. 7 Heracleum mantegazzianum; fig. 8 Kalanchoe laetivirens; fig. 9 Equisetum arvense; fig. 10 Impatiens glandulifera

fig. 8

Tiny plantlets that form on leaf margins fall to the ground and root to make new plants

A single plant can produce more than 20,000 seeds

Pods eject their seeds with such force that they land metres away from the parent plant

Mother plants

fig. 9

Brittle fragments that drop off may root and form underground networks that give rise to new plants

Fragmentation and regeneration

Spreading locally
Ruderal and competitive plants have developed similar strategies that allow their offspring to conquer local areas. They may spread using rhizomes to form colonies, by producing vast quantities of seed, or by creating new plants from their stems and leaves.

fig. 7

High fertility

fig. 10

Exploding pods

ADAPTATIONS
CROP DOMESTICATION

For a crop to be adopted and domesticated by humans, its wild ancestors need a suite of traits that are either present, or can be encouraged through breeding. In the early stages of domestication, these traits are generally already present in the wild plant, but chance mutations may make a wild plant suddenly more viable and attractive as a crop. Beyond the main reason the plant is grown – fruits, oil, medicinal properties, or something else entirely – the desirable characteristics to enhance can include disease resistance, as well as ease of seed storage, germination, reproduction, and growth.

fig. 1 Ficus carica; *fig. 2* Zea mays; *fig. 3* Vanilla planifolia; *fig. 4* Malus domestica; *fig. 5* Triticum monococcum subsp. monococcum; *fig. 6* Oryza sativa; *fig. 7* Lactuca serriola; *fig. 8* Lactuca sativa var. capitata; *fig. 9* Fragaria vesca; *fig. 10* Fragaria × ananassa

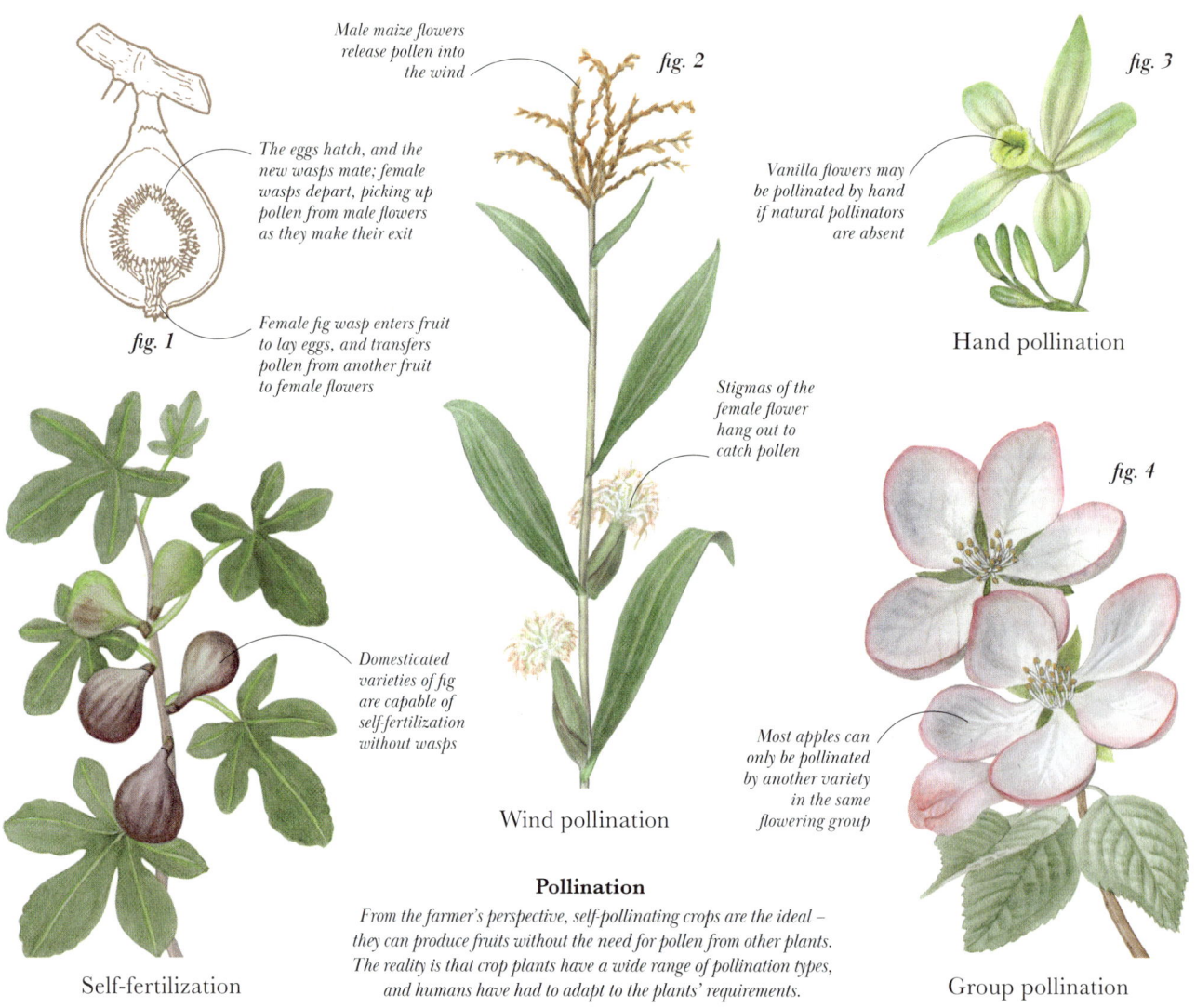

Pollination
From the farmer's perspective, self-pollinating crops are the ideal – they can produce fruits without the need for pollen from other plants. The reality is that crop plants have a wide range of pollination types, and humans have had to adapt to the plants' requirements.

Self-fertilization — Wind pollination — Hand pollination — Group pollination

• HUMAN ENVIRONMENTS •

fig. 5

Einkorn wheat

In domesticated wheat crops, the ear that holds the grain remains intact and does not split apart and drop the grains

Domestication

The domestication journeys of wheat and rice span millennia. Their grains are larger and easier to process than those of their wild ancestors. Several major sub-groups and thousands of varieties exist – from bread wheat to durum wheat, and basmati rice to Thai fragrant rice.

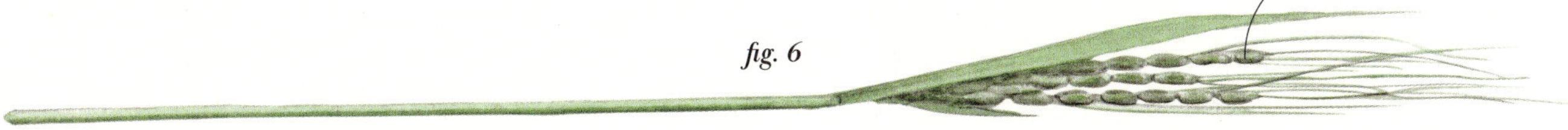

Basmati grains have been selected to be relatively long

fig. 6

Wild rice

fig. 7

Tough, slender leaves contain a bitter, milky latex and small spines

Wild lettuce

fig. 8

Large, soft leaves without latex no longer taste bitter

Cultivated lettuce

Changing character traits

Domestication has profound effects on crops, and they may be quite unlike their wild relatives. In many cases they are unlikely to thrive and reproduce without human intervention.

Small, sweet fruits with seeds that are dispersed by animals in the wild

fig. 9 Wild strawberries

Huge fruits can only be grown with careful cultivation

fig. 10 Cultivated strawberries

PLANTS OF
URBAN AREAS

Plants and humans have complex interdependent relationships – especially in urban landscapes. Built-up areas are adorned with street trees and shrubs, while city gardens are filled with cultivated varieties. These plants often establish self-sustaining populations and become part of the local flora. In doing so, they can sometimes become invasive: cultivated varieties are often selected to be vigorous and non-native plants are often less vulnerable to predators, herbivores, and diseases in their new situations.

1/ Vision in violet
Jacaranda mimosifolia

Frost-free towns and cities throughout the world are suffused with the sight and scent of the blue jacaranda's long-lasting flowers. Tall clusters of blooms completely cover the tree canopy, creating a lilac-purple cloud; after flowering, the fallen petals form a striking purple carpet under the trees. In its more temperate wild habitats in South America, the tree blooms mostly in spring, but when planted in warmer climes, it blooms both in the dry season – when the tree is bare of foliage – and the wet season, when the delicate feathery leaves flush out.

Trumpet-shaped flowers range from lilac to violet in colour

1 /

URBAN AREAS

2/ Deadly defence
Ricinus communis

Castor-oil plant is a strikingly colourful member of the Euphorbia family. Grown as a source of castor oil, and now more usually as a vivid ornamental, it has naturalized in many hot climates. Like other species in the family, its seeds contain the extremely toxic protein ricin, which gives this vital part of the plant a robust defence against herbivores and pests. This is one of the reasons – along with its ability to establish itself quickly in challenging conditions – that it has become a successful weed.

2/

Large, star-shaped leaves with serrated edges

3/ Community minded
Acacia mearnsii
Originally native to Australia and Tasmania, black wattle sustains symbiotic relationships that have helped it naturalize worldwide. Its seeds have small protein-rich appendages called elaiosomes that are a food source for birds and ants. Birds consume the seeds whole, while ants carry them off to feed on the elaiosomes, but both assist with seed dispersal. Black wattle also partners with special nitrogen-fixing bacteria that attach to its roots. The plant provides these bacteria with energy, while they supply nitrogen in return, promoting plant growth.

4/ Toxic vigour
Lantana camara
Originally from Central and South America, this evergreen shrub is now regarded as invasive in many warm climate countries. Much of its success lies in its combination of vigour and toxicity. Flowers are borne all year round, and each plant may produce as many as 12,000 fruits. Toxic leaves, flowers, and unripe berries, meanwhile, deter any predators.

5/ Chance hybrid
Senecio cambrensis
Inhabiting waste ground in north Wales and Scotland, the Welsh ragwort is one of several plants known to have evolved naturally in recent times. It formed when two different species, *Senecio vulgaris* and *S. squalidus*, interbred, creating a hybrid that was originally sterile. A chance genetic event later doubled the chromosomes in this hybrid and produced fertile plants – the new species *S. cambrensis*. Rare hybridizations like this are responsible for the huge species diversity in flowering plants.

6/ Acidic aroma
Ginkgo biloba
The sole surviving species in a genus native to China, ginkgo is now widely grown in cities in temperate regions. Its name comes from the Japanese "ginkyo" ("silver apricot") in reference to its fruitlike seeds. When these seeds fall to the ground and their fleshy, outer layer decays, it releases a smell reminiscent of rancid butter. This foul odour attracts scavenging animals, which ingest and disperse the seeds.

7/ Butterfly magnet
Buddleja davidii
Introduced to temperate gardens worldwide from its native China, buddleja has naturalized widely – and has been designated invasive in some areas. This talent for colonization stems from its ability to produce vast amounts of seed – a single plant can produce up to three million seeds each year – and to grow in challenging conditions. It thrives on derelict sites and the high walls of buildings because mortar creates the somewhat alkaline, well-drained conditions that replicate its natural habitat of Chinese riverbanks and limestone outcrops. Once established, it produces clusters of scented lilac flowers that attract a profusion of butterflies.

8/ Touch-sensitive
Berberis vulgaris
Originally from central and southern Europe, northern Africa and western Asia, the barberry is now a popular garden shrub. Its flowers have unusual stamens that are sensitive to touch. When an insect brushes against the base of a stamen while collecting nectar, the stamen snaps inwards and dabs the visitor with pollen. The mechanism involves specialized motor cells and electrical signals, and is similar to a knee-jerk response in humans. After a short time, the stamen returns to its original position, ready for another visiting insect.

8 /

Flowers develop into red, oblong berries

9/ Villainous reputation
Rhododendron ponticum

A hybrid of possibly three species, this hardy, vigorous rhododendron is native to the Iberian peninsula and the Black Sea region but has been introduced widely in gardens for the aesthetic appeal of its flowers. Its resilience, prolific seeding capability, and toxic leaves that deter herbivores, mean it has taken over urban green spaces and unmanaged woodlands. It forms a closed canopy that prevents native species from accessing sunlight, as well as hosting a disease that is deadly to trees.

10/ Clever clones
Taraxacum officinale

From northern temperate regions, dandelions have spread worldwide. Unusually for flowering plants, many dandelion populations reproduce without the need for pollination or fertilization. Instead, their tiny seeds – which are clustered into ball-shaped fruit-heads called "clocks" – are clones. Caught by the wind, these seeds are carried on hairlike parachutes to new sites, where they produce identical plants.

11/ Beauteous bracts
Bougainvillea spectabilis

Strongly associated with Mediterranean towns, this colourful, rapidly growing climbing vine is actually from South America. It appears to have large, striking flowers with bright purple, pink, or red petals, but these are bracts, specially adapted to attract the attention of pollinators such as butterflies and moths. The true flowers, which are clustered deep within the bracts, are tiny, white, and tubular.

12/ Fruitful spread
Rubus armeniacus

Known as the Himalayan blackberry – despite originating in Armenia and northern Iran – this plant has naturalized in many temperate areas. While it is prized for its sweet, edible fruit, similar to a large blackberry, its hardiness and prolific growth habit has made it an invasive species. Its canes can grow up to 5 m (15 ft) high, and form dense, prickly thickets that stifle native plants. The canes can also root from their tips, which provides the plant with an additional means of reproduction on top of its seeds, and makes it difficult to eradicate.

13/ Spontaneous species
Gunnera × cryptica

This stunning waterside plant is a huge, tough, and very new species. It originated when the large *Gunnera manicata* from Brazil and the cold-hardy *G. tinctoria* from Chile and Argentina were grown near each other in botanical gardens in the late 19th century. A spontaneous hybrid resulted that inherited the vigour of one parent and the resilience of the other, which made it so successful that it is now considered invasive in many countries.

14/ Pollution buster
Platanus × hispanica

A hybrid of plane trees from America and western Asia, the London plane is common in cities where it thrives due to its resistance to pollution. Its self-cleaning bark peels off, shedding layers that have trapped pollutants, while its leaves have a protective waxy coat.

15/ Secret weapon
Ailanthus altissima

Originally from China, tree of heaven is a vigorous deciduous tree widely grown in Mediterranean and similar climates. It is a pioneer of waste ground that spreads quickly through offshoots from its rhizomes, which can make it particularly successful in poor and disturbed soil in urban settings. To give it a competitive edge, it also releases toxins called allelochemicals into the environment to inhibit the growth of other species nearby.

9/

Bell-shaped flowers grow in large terminal clusters

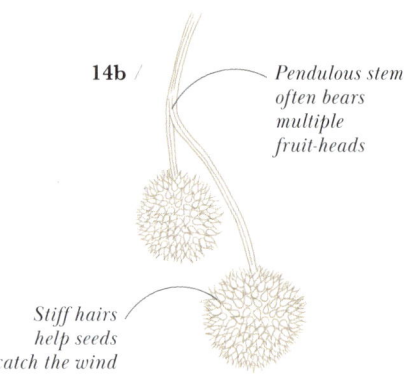

14b/

Pendulous stem often bears multiple fruit-heads

Stiff hairs help seeds catch the wind

Spiky spheres
The curious balls of softly spiky fruits are dense clusters of plane seeds, each one with a stiff hair. These clusters break apart in spring, spreading seeds.

• URBAN AREAS •

10 /

13 /

15 /

11 /

12 /

14a /

Seeds within the hull contain around 30 per cent oil

Oryza sativa
Rice is a semi-aquatic grass that is usually grown in shallow water or seasonally flooded areas.

6 *Colonial powers take maize and sunflower from the Americas to Eurasia around 500 years ago.*

5 *Colonial powers carry wheat to the Americas and Australia from the 16th century.*

2 *Probably cultivated in Central America about 9,000 years ago, maize spreads to North and South America.*

Helianthus annuus
The sunflower was domesticated in North America around 4,000 years ago and is well-suited to areas with warm, dry summers. A single head of this annual plant may produce more than 1,000 oil-rich seeds.

CROPS

Arguably some of the most successful plants, crops are cultivated across the globe, and are commercially highly significant.

Most crops have ancient origins from wild ancestors, but today they are cultivated in monocultures, or single-crop farms, under relatively controlled conditions. This means that they can face huge challenges from pests and environmental issues such as climate shifts. A look back at their wild forerunners offers potential for the future as these plants are often more genetically adaptable to changing conditions.

HISTORIC MIGRATION

Amaranthus caudatus →	*Helianthus annuus* →	*Oryza sativa* →
Sorghum →	*Triticum* →	*Zea mays* →

AREAS OF MODERN CULTIVATION

- *Amaranthus caudatus*
- *Helianthus annuus*
- *Oryza sativa*
- *Sorghum bicolor*
- *Triticum aestivum*
- *Zea mays*

Pendulous flower head develops fruit with a grainlike seed

Amaranthus caudatus
First domesticated around 5,000 years ago, amaranth is a salt-tolerant, mid-altitude food crop of the Andes. It is now becoming a popular crop in North America and Europe.

• HUMAN ENVIRONMENTS •

Protective fibrous husk encloses each grain of rice

Zea mays
Maize is a robust grass with an efficient system of C4 photosynthesis that optimizes the use of carbon dioxide. As a result, it grows speedily in hot climates, yielding a large, starch-rich grain.

❶ Domesticated more than 10,000 years ago, wheat spreads from the Fertile Crescent in western Asia to Europe, northern Africa, and across Asia.

❸ Rice is first domesticated in central China's river basins around 8,000 years ago. It spreads to Southeast Asia and India, where a local subspecies is adopted shortly after. Rice is introduced to the Americas in colonial times.

Colourful cultivars of maize are grown on a small scale

❹ Sorghum is cultivated in sub-Saharan Sahel habitats about 6,000 years ago, and moves southwards.

Thick, waxy coat on leaves and stems reduces water loss

Stems on traditional varieties of wheat were long and used in thatching, while modern ones are shorter

Triticum aestivum
The most widely grown grain, wheat thrives in temperate areas. The ears of grain on cultivated forms remain intact rather than scattering their seed, so it relies on human intervention for its spread.

Sorghum bicolor
A native of Africa, this high-yielding nutritious grain uses the efficient C4 form of photosynthesis, and tolerates relatively arid conditions.

Banana

Musa acuminata

The banana and its relatives – savoury plantains and enset – have been domesticated over thousands of years and are staple crops around the world. The most widely grown variety is the dwarf Cavendish (*Musa acuminata*), which is cultivated in the Caribbean, Madagascar, Australia, and South and Southeast Asia. The species was originally recognized as *M. cavendishii* from specimens grown at Chatsworth House in Derbyshire, UK, by renowned horticulturist and architect Joseph Paxton. They were propagated from a shipment from Mauritius received by William Cavendish, 6th Duke of Devonshire, who owned Chatsworth. The species now represents almost 100 per cent of global banana exports. It is considered particularly desirable for consumers because it is a triploid – it cannot produce seeds – so it is cultivated via clones, giving a reliable flavour.

In addition to being a popular food, banana fruits are used in traditional medicine to treat everything from diabetes to fungal skin complaints – with varying degrees of success. Eating the potassium-rich fruits is also widely thought to alleviate muscle cramp. Some Asian and African communities use the flowers in traditional medicine to treat bronchitis, dysentery, and ulcers.

Bract peels back to expose the flowers

Leathery bract protects unopened flowers

Leaf paper
Banana leaves are used to serve and wrap food in many parts of Asia. They have even been used as a writing surface to practice calligraphy, as shown in this nineteenth century painting by Qian Hui'an.

Stout "stems" made of leaf sheaths contain air-cells, that make them light and strong

Flowers are produced in clusters, which mature into bunches – or "hands" – of fruit

"The flavour, when in perfection, combines that of the pine-apple, the melon and the pear."

Joseph Paxton, "On the Culture of the Musa Cavendishii, as practiced at Chatsworth", 1837

PLANTS OF
AGRICULTURAL AREAS

Plants that are cultivated have all been chosen because their natural adaptations provide benefits to humans. In many cases these domesticated plants have been enhanced through selection and breeding to emphasize the key qualities for which they are used, sometimes resulting in plants that are quite unlike their wild relatives. These techniques have made domesticated species incredibly successful in terms of yield and ease of cultivation, allowing them to occupy far more space than their wild relatives might ever have achieved on their own.

1/ Glorious gourds
Cucumis melo

The swollen fruits of squashes and melons were probably an adaptation that arose to allow seeds to be dispersed by large mammals and birds. The animals would break through the tough rind and eat the flesh inside, inadvertently scattering the seeds; any seeds gulped down were dispersed in their droppings. Although a few *Cucumis* species contain compounds that are toxic to humans, numerous other species have been domesticated independently on different continents, giving testimony to their delicious versatility.

1a /

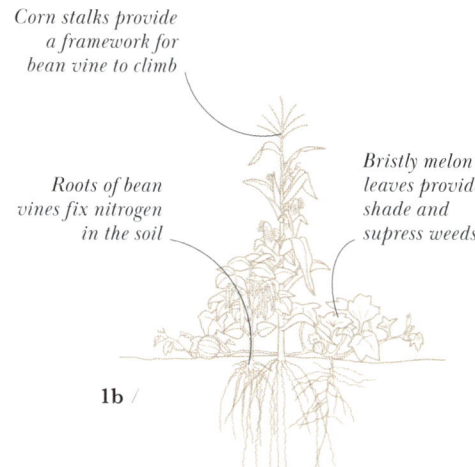

Corn stalks provide a framework for bean vine to climb

Roots of bean vines fix nitrogen in the soil

Bristly melon leaves provide shade and supress weeds

1b /

Companion planting
The "three sisters" method of companion planting groups corn, beans, and a squash or melon together. Sown in the right sequence – corn first, then the beans, and finally the squash – the plants each bring benefits for all.

• AGRICULTURAL AREAS •

2/

4/

5/

3/

6/

2/ At home in the water
Oryza sativa

"Wild rice" is a catch-all term for a range of grasses with slender grains that often like wet conditions. Asian rice (*Oryza sativa*) and African rice (*O. glaberrima*) were independently domesticated from wild species that thrive in marshlands inundated with water. Both are able to cope with waterlogged soils through adaptations such as aerenchyma – air spaces within the rhizomes and shoots. Asian rice is broadly split into the sticky, short-grained *Japonica* rice and long-grained non-sticky *Indica* variety.

3/ Mediterranean marvel
Vitis vinifera

Wild grapes were a staple of hunter-gatherer diets for millennia, but they were only domesticated around 7,000 years ago. Now, there are over 5,000 cultivars of the common grape vine. Unlike the separate male and female plants of its wild ancestor *Vitis sylvestris*, the common grape vine is hermaphroditic, so it is able to self-pollinate. This has given it the advantage of bearing more consistent and predictable fruit.

4/ Soy joy
Glycine soja

The wild soybean is a versatile and easily grown species partly because it produces fairly large seeds that can remain dormant and germinate readily in challenging conditions. Its symbiotic relationship with rhizobia bacteria allows it to capture nitrogen. The wild soybean provides the bacteria with a home and food, and the bacteria convert atmospheric nitrogen into ammonia, which the plant uses to build proteins that are vital for growth. First domesticated in China, the cultivated soybean, is one of the world's most-grown legume crops.

5/ Tower of strength
Phyllostachys edulis

Bamboo can grow at a rate of 114.5 cm (45 in) a day, which makes it an incredibly sustainable crop. Their strong, segmented stems reach heights of up to 8 m (26 ft) and are extremely versatile, used for everything from scaffolding poles to lightweight water bottles. Its rapidly spreading underground rhizome system remains intact after stems are harvested, and sends up new shoots for the next season.

6/ Back-up plan
Solanum tuberosum

The potato is a hugely diverse crop with many wild relatives still grown in its Andean homeland. The plants cultivated worldwide are from a limited genetic stock, which makes them susceptible to disease, but their wild relatives offer a key source of genetic diversity to help protect this popular crop into the future.

7 /

8 /

9 /

10 /

11 /

12 /

AGRICULTURAL AREAS

7/ Protector and painkiller
Papaver somniferum
Scratching the unripe seed pods of the opium poppy releases a milky-white substance (latex), which is rich in opiates including morphine and codeine. The latex functions as a protective toxin that deters herbivores from eating the plant. A combination of natural occurrence and human selection has resulted in the particularly high quantities of opiates found in the opium poppy. Their medicinal and hallucinogenic properties made opium poppies important ritual plants in early cultures, such as the Minoan civilization that inhabited Crete during the Bronze Age.

8/ Bounty of Eden
Malus domestica
With small, often sour-tasting fruits, wild species of apples are mostly eaten by birds and mammals that scatter the seeds in their droppings. The fruits are long-lasting, and well-suited for storage. Selection for larger, sweeter fruits began in central Asia, and resulted in thousands of varieties, many of which are lost or at risk of extinction.

9/ Chemical defences
Mangifera indica
An iconic tropical fruit tree, mango is related to poison ivy and poison oak, all of which contain urushiol, a colourless defensive chemical that deters pests and can trigger dermatitis in humans. Mango was domesticated around 5,000 years ago in Southeast Asia and southern India, resulting in two distinct populations. Mango trees typically live up to 60 years but some individuals can also be surprisingly long-lived, like the 300-year-old specimen in East Khandesh, India, which is still bearing fruit.

10/ Fragrant florals
Lathyrus sativus
This tough, drought-resistant grass pea is cultivated in Kenya and India, as well as parts of the Iberian peninsula. It is grown mainly as animal fodder and is so resilient that it can survive where drought has killed off other crops. The seeds are used in traditional foods in North Africa, western Asia, and southern Europe, but may be toxic in large quantities. In some varieties the toxins have been bred out.

11/ Space-hungry palm
Elaeis guineensis
The plump fruits of the oil palm ripen through a series of bright shades to deep red or black. Such shiny, contrasting colours are attractive to birds including the grey parrot. The genus name, *Elaeis*, refers to the abundant oil in the nutritious kernel that led to its domestication. Native to west and central Africa, it is now grown throughout the wet tropics – often at the expense of biodiverse forest.

12/ Pick of the bunch
Musa spp.
This diverse genus of flowering plants includes bananas and plantains. Although they can grow as high as trees, banana and plantain "trees" are really gigantic herbs. Their "trunk", or pseudostem, is made up of tightly rolled layers of leaf sheaths rather than wood. Once the "tree" bears a bunch of fruit, made up of tiers, or "hands", the pseudostem dies back, but is replaced by new suckers from its underground rhizome. The suckers grow into a pseudostem and the fruiting cycle continues.

13/ Lucky break
Ficus carica
Wild figs are usually pollinated by tiny, highly specialized wasps – an example of a co-evolutionary partnership. Over 11,000 years ago, humans made a chance find of self-pollinating mutants of the wild fig and propagated them, creating the edible fig. Not needing a pollinator proved to be both more convenient and productive, and resulted in the spread of the edible fig throughout the Mediterranean and beyond.

Fruit is an inverted inflorescence – the flowers bloom inside the fleshy structure

13/

14/ Heady scent
Rosa × damascena

The damask rose is one of hundreds of species in the genus but is particularly renowned for the sweet scent, which acts as a long-distance signal to attract its insect pollinators. Humans have used the petals to produce rose oil and rose water since at least the 10th century. Like many cultivated species, damask rose is a hybrid of others, with at least *Rosa gallica* and *R. moschata* (the musk rose) contributing to its parentage.

15/ Global fibre
Gossypium hirsutum

Several species of cotton have been domesticated completely independently across separate continents for the fluffy fibres that protect their seeds and aid their dispersal on the wind. *Gossypium hirsutum*, also called upland cotton, is the most planted species of cotton and accounts for approximately 90 per cent of world cotton production. After its cup-shaped yellow or white flowers wither, a green boll forms, which bursts open, to reveal the fluffy fibres inside.

16/ Easy harvest
Triticum monococcum

The discovery of einkorn wheat was critical in the development of domesticated wheat. Wild grasses often drop their individual grains separately to ensure better dispersal, but on einkorn the whole head remains intact, which makes it easy for farmers to harvest. Einkorn is considered an "ancient grain", largely unchanged for thousands of years. Its gluten structure is simpler than that of domesticated wheat, so it is easier to digest.

17/ Energy-rich meal
Olea europaea

The large black fruits of olive trees are single-seeded and oil rich. This made their wild relatives an ideal food for birds preparing for winter or migration and the birds would pay back the service by naturally dispersing the seeds. These wild relatives – sometimes treated as subspecies or a distinct species – are referred to as "oleaster" and along with the domesticated species (*Olea europaea* subsp. *europaea*) are found throughout the Mediterranean region.

18/ South Pacific staple
Colocasia esculenta

The starchy rhizome of taro provides energy for the large leaves and striking flowerheads of this huge herb. They can be dug up, stored, and replanted after weeks – a benefit that made it a staple food during the Polynesian expansion throughout the Pacific Islands. Its rhizomes contain toxic, spiky calcium oxalate crystals to protect them from herbivores, but techniques such as physically bashing the rhizome and roasting it break down the crystals and the toxins inside to make taro edible.

19/ Fuel of ancient empires
Cicer arietinum

The seeds of the humble chickpea are wonderfully nutritious. As members of the legume family, their ability to capture their own nitrogen means that they are packed with protein. Among the earliest crops domesticated in the Fertile Crescent of western Asia, they are commonly found in archaeological remains of ancient empires from the Egyptians to the Hittites and remain a staple food for millions of people today.

20/ Sweet seeds
Cocos nucifera

The giant seed of the coconut contains both white flesh and a liquid endosperm – coconut water – which supplies the energy required for germination. First domesticated in maritime Southeast Asia, the coconut proved a very versatile plant. Its flesh is sweet and rich in fat. Its tough seed coat also doubles as a useful drinking vessel, and its coir – the fibrous husk of the fruit – has been used in everything from matting to peat-free composts.

Double flowers feature more petals and fewer stamens, so are less accessible to pollinating insects

14/

Strong, curved thorns deter herbivores

• AGRICULTURAL AREAS •

15 /
16 /
17 /
18 /
19 /
20 /

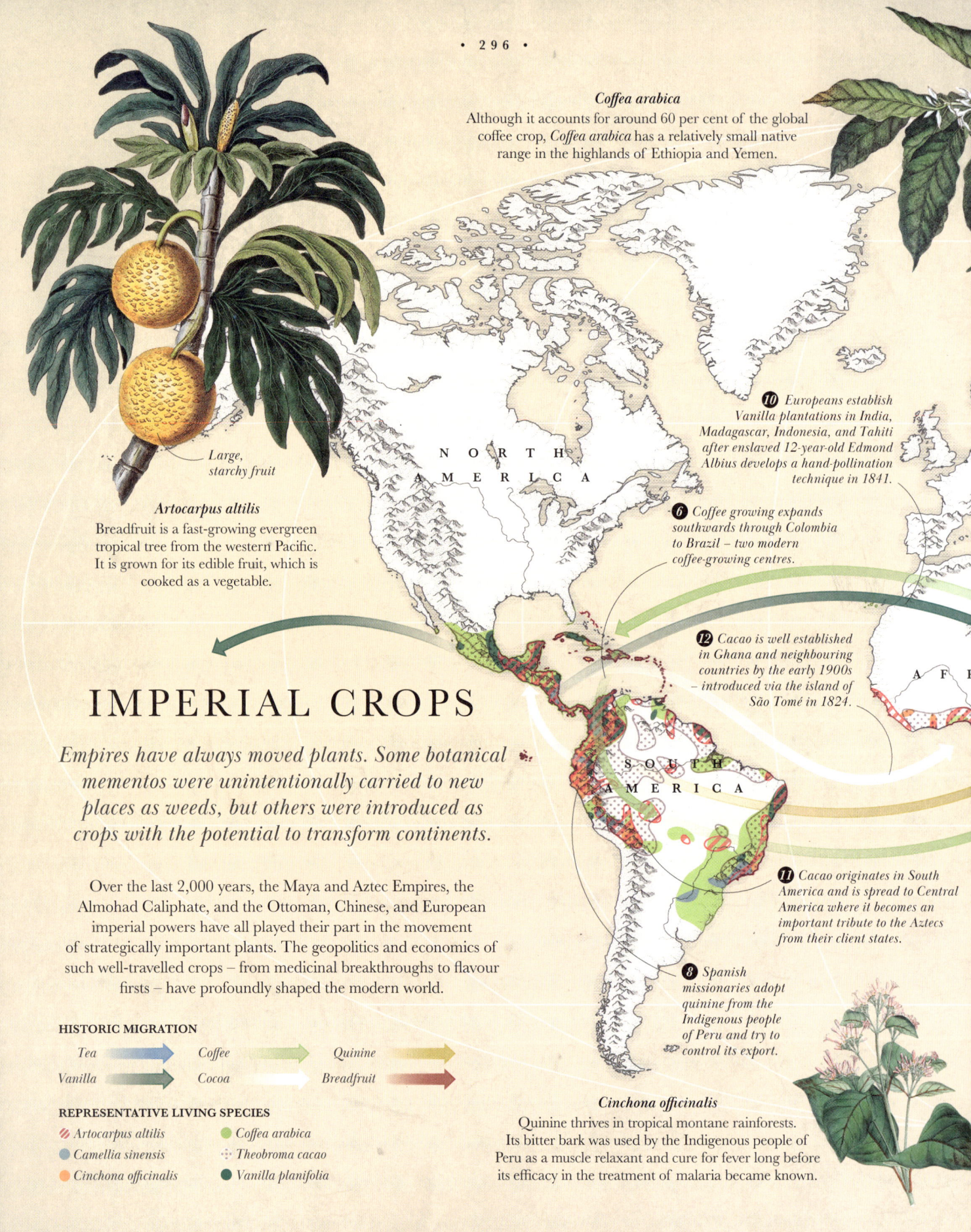

IMPERIAL CROPS

Empires have always moved plants. Some botanical mementos were unintentionally carried to new places as weeds, but others were introduced as crops with the potential to transform continents.

Over the last 2,000 years, the Maya and Aztec Empires, the Almohad Caliphate, and the Ottoman, Chinese, and European imperial powers have all played their part in the movement of strategically important plants. The geopolitics and economics of such well-travelled crops – from medicinal breakthroughs to flavour firsts – have profoundly shaped the modern world.

Coffea arabica
Although it accounts for around 60 per cent of the global coffee crop, *Coffea arabica* has a relatively small native range in the highlands of Ethiopia and Yemen.

Artocarpus altilis
Breadfruit is a fast-growing evergreen tropical tree from the western Pacific. It is grown for its edible fruit, which is cooked as a vegetable.

Large, starchy fruit

Cinchona officinalis
Quinine thrives in tropical montane rainforests. Its bitter bark was used by the Indigenous people of Peru as a muscle relaxant and cure for fever long before its efficacy in the treatment of malaria became known.

10 Europeans establish Vanilla plantations in India, Madagascar, Indonesia, and Tahiti after enslaved 12-year-old Edmond Albius develops a hand-pollination technique in 1841.

6 Coffee growing expands southwards through Colombia to Brazil – two modern coffee-growing centres.

12 Cacao is well established in Ghana and neighbouring countries by the early 1900s – introduced via the island of São Tomé in 1824.

11 Cacao originates in South America and is spread to Central America where it becomes an important tribute to the Aztecs from their client states.

8 Spanish missionaries adopt quinine from the Indigenous people of Peru and try to control its export.

HISTORIC MIGRATION
Tea → Coffee → Quinine →
Vanilla → Cocoa → Breadfruit →

REPRESENTATIVE LIVING SPECIES
- *Artocarpus altilis*
- *Camellia sinensis*
- *Cinchona officinalis*
- *Coffea arabica*
- *Theobroma cacao*
- *Vanilla planifolia*

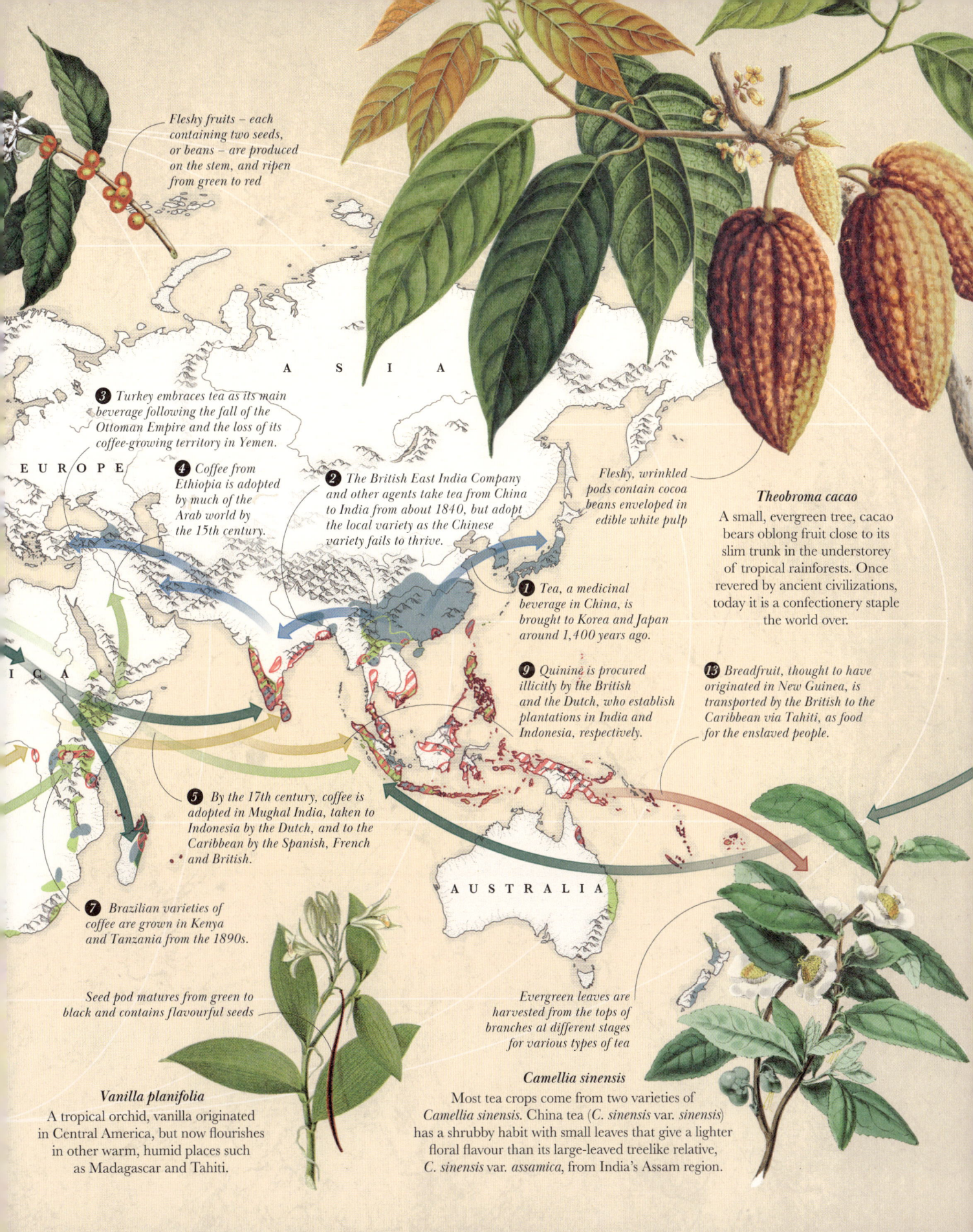

• HUMAN ENVIRONMENTS •

POLLUTION

Humans have profoundly changed Earth's seas, soils, and skies. The use of fossil fuels, industry, and many agricultural practices have altered climate and nutrient cycles, and polluted terrestrial ecosystems, waterways, and the atmosphere. Although aerial pollution is a global issue, its effects are particularly pronounced in urban areas with high levels of emissions from vehicles.

Pollutants and particulates can have a significant effect on plants, but resourceful species in urban areas have found ways to survive – many weedy species, for example, can thrive in polluted areas where they benefit from reduced competition and the availability of plenty of nitrogen. Humans have adopted a broad palette of resilient plants to grow in urban settings, including street trees such as ginkgo that can tolerate compacted soils, hard street surfaces above their roots, and an array of atmospheric and soil pollutants.

Although ginkgo is often a feature of city streets, there was a time when its wild population was in danger of going extinct – it was preserved through its extensive use in China and Japan to decorate sacred temples. This beautiful, long-lived tree has earned a place in the urban flora due to a number of specific adaptations that make it remarkably tolerant of pollution. It can regulate how much harmful gas it takes in by reducing the size of its stomatal opening – adjustable pores on the leaves that act like biomechanical valves. Inside, the leaves have a thick middle layer of cells (the mesophyll), which disperse and dilute the concentration of pollutants. The tree is also able to sequester certain pollutants within its leaf wax and tissues without suffering significant harm.

Ginkgo as a genus has survived multiple extinction events during its more than 200-million year history in the fossil record; its resilience is thought to be the reason why it was one of the few tree species to survive the nuclear bomb at Hiroshima in 1945. Today, its offspring graces botanic gardens around the world as a powerful symbol of hope and peace.

Spectacular sequesterer
This more than 1,000-year-old ginkgo in Chengdu, China, helps to clean the air by accumulating pollutants in its delicate-looking foliage. Its leaves are shed harmlessly when the temperature drops in autumn, creating a colourful flaming yellow carpet on the city street.

PLANTS OF
PROTECTED AREAS

Geographically defined spaces that are managed for long-term conservation harbour a wealth of plant life across the globe. Some of these protected areas were specifically created to provide vital support to endangered trees, such as the Monkey Puzzle (*Araucaria araucana*) in Chile's Conguillío National Park. These sites come in all shapes and sizes – from local nature reserves to vast swathes of land – but are often scattered, lacking interconnectedness, and as human populations and pressures increase, the need for them has never been greater.

Crown of tree forms distinctive candelabra shape

1 /

1 / Precious pine
Araucaria rulei

The graceful pine is endemic to the South Pacific island of New Caledonia where it grows on nickel-rich soils. These environments are toxic to most other plants, but the graceful pine thrives on soils that few species can tolerate. The creation of open-cast nickel mines on its natural habitat has caused numbers to suffer a severe decline. The outlook for the graceful pine looks more positive, however, after the formation of a forest restoration programme in New Caledonia, which plans to increase the number of protected areas in this unique "island ark".

• PROTECTED AREAS •

2/

3/

4/

2/ Bell-shaped beauty
Abutilon menziesii
Ko'oloa'ula, as its known in its native habitat, is an endangered flowering shrub only found on four Hawaiian islands. It is restricted to scattered spots in the wild, but its beautiful flowers – ranging in colour from pink to deep red – mean that it has been planted in dry areas across the islands. The open bell shape of its flowers funnels pollinators, especially bees, deep inside to access the nectar, where they make contact with the reproductive organs, and transfer pollen. Although it faces threats such as herbivory, competition with invasive plants and, increasingly, wildfires, conservation efforts are showing promising signs of restoring the plants and their habitats.

3/ One of a kind
Encephalartos woodii
Habitat loss and overcollection led to the extinction of all but one example of this palm-like woody plant and it can be found at the Royal Botanic Gardens Kew in London. Like other cycads, this species has separate male and female plants and the lone survivor is a male. It does grow "pups" – separate stems that can be removed and grown elsewhere – but these are all genetically identical males.

4/ Coastal icon
Sequoia sempervirens
The California or coastal redwood is the world's tallest tree, reaching heights over 115 m (380 ft). It has an extraordinary lifespan, with some trees living for thousands of years. They are aided in this accomplishment by their vital relationship with sea mist, which supplies them with moisture during dry periods in the summer. The breathtaking scale and majesty of these trees in such a defining landscape inspired Colonel George Stewart and John Muir, among other pioneering conservationists, to establish a national park – the Sequoia National Park in California – to preserve these iconic conifers.

5/ Seasoned survivor
Begonia socotrana
This cliff-dwelling plant is endemic to a small part of the Hajhir mountains on the island of Socotra in the Indian Ocean. Its bulbils and succulent stems help it to store and retain water, allowing it to survive during prolonged dry spells. It is also a winter-flowering species, so it avoids the very hottest months on the arid, semidesert island, when pollinators are few and far between.

5/

Distinctive round leaf with central stalk

6/ Living fossil
Metasequoia glyptostroboides
The Dawn redwood is an endangered conifer found in a few locations in central China. It was known only from its fossil record and believed to be extinct until living specimens were discovered in a remote valley by a botanist in the 1940s. The dawn redwood is the only living member of its genus, and unlike the related *Sequoia* genus, its needles are arranged in opposite pairs along its branchlets rather than in a spiral. It is one of the few conifers to lose its needles in autumn, after turning striking shades of reddish-orange and coppery brown.

7/ Miniature mountain moss
Herbertus borealis
The delicate little Northern prongwort is a moss-like leafy liverwort found in a single nature reserve in the mountains of northwestern Scotland. It forms extensive yellowish, orange, or brown mats that can grow up to a height of 20 cm (8 in). The purpose of the hook-like prongs is not yet known, but it could be to help prevent the plant from drying out. Like many relatives, it can also enter a form of dormancy as a last resort in times of low moisture.

8/ Bottled water
Adansonia fony var. *rubrostipa*
The swollen trunks of baobabs are a striking sight in semi-arid and arid regions of mainland Africa, Madagascar, and Australia. Found in Madagascar, the fony baobab is the smallest species. As with other baobabs, its bottle-shaped trunk, with its distinctive reddish brown bark, stores water in the soft woody interior. As the tree ages, the trunk can die from the inside, leaving hollows, crevices and caverns, which provide shelter to a wealth of other organisms.

9/ Rewarding reptiles
Nesocodon mauritianus
The extraordinary blood-red colour of the bloody bellflower's nectar inspired its common name. This nectar is a reward for the large, vividly coloured day geckos that are thought to pollinate the plant – hence the particularly large opening to the flower. These geckos, like the plant, are endemic to the island of Mauritius.

10/ Record-breaking flower
Rafflesia arnoldii
The colossal bloom of the stinking corpse lily appears to erupt from the forest floor on the islands of Sumatra and Borneo. It is a parasitic plant that is entirely dependent on its host, *Tetrastigma* lianas, for all its nutrition and water. The flower – which can measure 1 m (3 ft) across and only appears for around a week – is the single largest individual flower in the world. It is thought to both look and smell like a decomposing animal in order to draw in scavenging flies as pollinators.

11/ Spectacular stench
Amorphophallus titanum
Growing up to 3 m (10 ft) tall, the titan arum possesses one of the most spectacular flower heads of any plant. The underground tuber that fuels this giant inflorescence builds up resources for several years through a single vast leaf that resembles a small tree. Only once it has stored up enough energy does the plant bloom. The flower spike, or spadix, generates its own heat and releases a scent reminiscent of rotting flesh that draws in carrion-feeding pollinators. This gives this plant the English name "corpse flower".

12/ Fungus imitator
Dracula spp.
The genus name Dracula translates as "little dragon" and reflects the dragon-like appearance that the long sepal spurs give the flowers rather than having anything to do with the legendary vampire. Inhabiting the forests of central America and the northwest Andes, the plants are epiphytic, growing on other plants. Their pendulous flowers attract specific pollinators, such as fruit flies, by mimicking fungi. Some species give off fungal-like scents and their labellum, or liplike landing pads, have thin, bladelike "gills" like those of fungi.

Spotted, triangular flower with long, tail-like spurs

12/

13 / Lasting legacy
Nesiota elliptica
Native to the cloud forest of the island of Saint Helena, the last Saint Helena olive died out in 1994 – a victim of grazing by introduced goats and climate change. Confusingly, it was not related to the olive, but to the buckthorn family. In 1996, the Peaks National Park was established on the island to protect its cloud forest and remaining native species. The final cultivated Saint Helena olive perished in 2003, but several specimens were dried and preserved to serve as a vital genetic record.

14 / Early riser
Iris koreana
This rare little woodland iris is endemic to the Korean peninsula. It forms small clumps that are connected by slender rhizomes, which allows it to spread quickly. Like many other forest floor species, it tends to flower before the leaves of the surrounding trees emerge and the spring canopy closes over. As such, it attracts early emerging insect pollinators, like bumblebee queens and solitary bees.

15 / Strongly scented
Lathyrus odoratus
The sweet pea is originally from southern Italy and Sicily. Although a profusion of cultivated varieties exist, the wild species is considered critically endangered in the wild due to overcollection. The sweet pea is mostly self-pollinating but nevertheless exudes a strong fragrance that may attract additional pollinators. This enables cross-pollination and enhanced genetic diversity, giving the population resilience.

16 / Flash of genius
Dipteryx oleifera
Known as the tonka bean, this lofty tree of South American rainforests, is an "emergent", sitting high above the canopy of others. It may have developed a remarkable strategy for killing off the competition posed by climbing plants like lianas, by growing to unusual heights it receives a number of lightning strikes. The tonka bean is relatively unscathed when it is hit by lightning, but lianas and neighbouring trees may be damaged or killed.

Small pink flowers sit on branched inflorescences

13 /

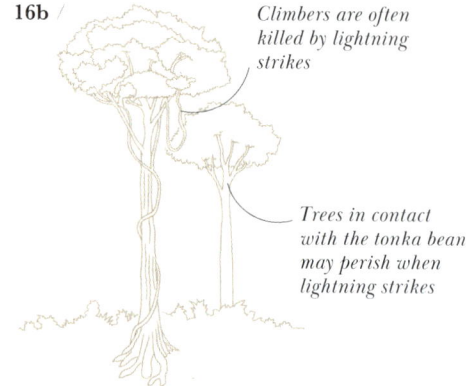

16b /

Climbers are often killed by lightning strikes

Trees in contact with the tonka bean may perish when lightning strikes

Clearing out the competition
The tonka bean has high internal conductivity, so it can safely dissipate lightning strikes and transfer the energy to competing neighbouring trees.

17 / Up in the trees
Rhododendron himantodes
This dwarf rhododendron with its narrow, needle-like leaves and distinctive red stamens is usually found as an epiphyte, growing on larger trees in its Borneo cloud forest habitat. Although this strategy allows it better access to light, rooting in cracks on the bark of other plants can make accessing water difficult, so it has developed shiny, leathery leaves to prevent water loss. A golden-brown scaly covering on its stems, buds, and white flowers, prevents insect herbivores from eating the leaves and may also help the plant to retain water.

18 / Canyon conifer
Wollemia nobilis
This critically endangered evergreen conifer was only scientifically known from fossils until a chance discovery in a small canyon in Australia's Blue Mountains in 1994. Around 80 trees are at this site, which is now carefully managed, but a programme of propagation and translocation to safe sites such as botanical gardens has shown some promising results.

19 / Wheel of death
Aldrovanda vesiculosa
Water wheel is a fully aquatic predatory plant found in scattered spots around the world. Like its relative the Venus flytrap (*Dionaea muscipula*), water wheel uses modified leaves as traps to catch aquatic creatures. Trigger hairs lining the edge of the trap set off a rapid midrib movement that causes the trap to snap shut – ten times faster than the Venus flytrap.

• PROTECTED AREAS •

GLOSSARY

Aerial root
A root that grows from the stem of a plant that is located above the ground.

Algae
Simple, flowerless, mainly aquatic plantlike organisms that contain the green pigment chlorophyll but do not have true stems, roots, leaves, or vascular tissue.

Angiosperm
A flowering plant that bears ovules, later seeds, enclosed in ovaries. Angiosperms are divided into three main groups for classification: early-diverging angiosperms, monocots, and eudicots. *See also* Early-diverging angiosperm, Monocot, Eudicot.

Annual
A plant that completes its entire life cycle – germination, flowering, seeding, and dying – in one growing season.

Anther
The part of a flower's stamen that produces pollen; it is usually borne on a filament.

Anthocyanin
Plant pigment molecules that are responsible for red, blue, and purple colours in leaves, shoots and in flowers.

Aril
A berrylike, fleshy, hairy, or spongy layer around some seeds.

Axil
The upper angle between a stem and a leaf, where an axillary bud develops.

Axillary bud
A bud that develops in the axil of a leaf.

Bark
The tough covering on woody roots, trunks, and branches.

Berry
A fruit with soft, juicy flesh surrounding seeds that have developed from a single plant ovary.

Biennial
A plant that flowers and dies in the second growing season after germination.

Bipinnate
A compound leaf whose leaflets are divided into yet smaller leaflets.

Bract
A leaf that has modified into an attractive or protective structure around the base of a flower or flower cluster that often protects young buds. Some bracts are large, brightly coloured, and resemble flower petals to attract beneficial insects, while others look like leaves, although they may be smaller and shaped differently from the other leaves on the plant.

Broadleaf
Describes flowering trees and shrubs that have broad, flat, sometimes deciduous leaves, rather than the narrow, needlelike leaves of conifers.

Bud
An immature organ enclosing an embryonic branch, leaf, inflorescence, or flower.

Bulb
A modified underground bud that acts as a storage organ. It consists of one or more buds and layers of swollen, colourless, fleshy scale leaves on a shortened, disclike stem.

Bulbil
A small, bulblike organ, often borne in a leaf axil, occasionally on a stem, leaf margin, or on a flower head. Bulbils often drop off and form new clonal plants.

Calyx (pl. Calyces)
The outer part of a flower, formed from a ring of sepals, that is sometimes showy and brightly coloured but usually small and green. The calyx forms a cover that encloses the petals while in bud.

Cambium
A layer of tissue that can produce new cells that increase the girth of stems and roots.

Canopy
The layer of high-level foliage found in a forest and formed by the crowns of individual trees.

Capsule
A dry fruit containing many seeds that has developed from an ovary formed from two or more carpels. It splits open when ripe to release its seeds.

Carotenoid
Plant pigment molecules responsible for yellow and orange hues.

Carpel
The female reproductive part of a flower that consists of an ovary, stigma, and style. *See also* Ovary, Pistil.

Cauliflorous
A term used to describe flowers and fruits that develop directly on a tree's trunk or branches rather than at the ends of its twigs.

Chlorophyll
The green pigment inside plant cells that allows leaves and sometimes stems to absorb light and carry out photosynthesis.

Chloroplast
The structures inside plant cells that contain chlorophyll, where starch is formed during photosynthesis.

Cladode
A modified stem that both resembles and performs the function of a leaf.

Climber
A plant that grows up a vertical surface, such as a rock or a tree, using it for support. Climbers do not gain nourishment from their plant hosts, but may weaken them by blocking out light and competing with them for moisture. *See also* Liana.

Compound leaf
A leaf composed of two or more similar parts. *See also* Leaf, Leaflet.

• GLOSSARY •

Cone
The densely clustered bracts of conifers and some flowering plants that may develop into a woody, seed-bearing structure, such as a pine cone. See also Bract.

Conifer
An evergreen tree or shrub that usually has needlelike leaves and naked seeds that develop on scales inside cones.

Corm
A bulblike underground swollen stem or stem base, often surrounded by a papery tunic. See also Rhizome, Tuber.

Corolla
A ring of petals on a flower.

Cortex
The region of tissue between the outer epidermis or bark of a plant and the central vascular cylinder.

Corymb
A broad, flat-topped, or domed inflorescence of stalked flowers or flower heads arising at different levels on alternate sides of a stem.

Cotyledon
A seed leaf that acts as a food store or unfurls shortly after germination to fuel plant growth.

Cross-fertilization
The fertilization of the ovules of a flower as a result of cross-pollination.

Cross-pollination
The transfer of pollen from the anthers of a flower on one plant to the stigma of a bloom on another plant. See also Self-pollination.

Culm
The jointed, usually hollow, flowering stem of a grass or bamboo.

Cultivar
A contraction of the term "cultivated variety", which is used to describe a plant that generally only exists as a result of human cultivation.

Cupule
A cup-shaped structure made of bracts joined together, as in an acorn. See also Bract.

Cuticle
A protective, waxy, water-repellent coating on the outer cells of the epidermis of some plants.

Cyme
A branched inflorescence, flat or round-topped, with each stem, or axis, ending in a flower, the oldest at the centre and the youngest arising in succession from the axils of secondary bracts (bracteoles).

Deciduous
Used to describe plants that shed their leaves at the end of a growing season and then renew them at the beginning of the next. Semi-deciduous plants lose only some of their leaves at the end of the growing season. See also Evergreen.

Dioecious
A plant that bears unisexual flowers, with male and female blooms occurring on separate plants. See also Hermaphrodite, Monoecious.

Drip tip
The tapering tip of a leaf or leaflet that helps to direct rainwater run-off.

Drupe
A fleshy fruit containing a seed with a hard coat. See also Endocarp.

Early-diverging angiosperm
A diverse group of flowering plants, also known as the basal angiosperms, that branched off relatively early in the evolutionary history of flowering plants. This group includes magnolias, nutmeg, black pepper, and bay laurel. See also Monocot, Eudicot.

Ecosystem
A collection of plant, fungal, and animal species living in the same habitat that form a complex interdependent relationship.

Emergent
Coming out of, or emerging from.

Endemic
A species that is native to a particular geographic area, such as an island, forest, mountain, or country, and does not naturally occur elsewhere.

Endocarp
The innermost layer of the pericarp of a fruit. See also Drupe, Pericarp.

Epidermis
The protective outer layer of cells in a plant.

Epiphyte
A plant that grows on the surface of other plants without being parasitic or stealing nutrients from its host; it often obtains moisture and nutrients from the atmosphere without rooting into soil, but may root in the organic matter that accumulates in crevices on the host plant.

Eudicot (Eudicotyledon)
A flowering plant that has two seed leaves, or cotyledons, including many plants formerly described as "dicots". Most eudicots have broad leaves with branching veins and the floral parts, such as petals and sepals, arranged in groups of four or five. See also Early-diverging angiosperm, Monocot.

Evergreen
Describes plants that retain their foliage for more than one growing season; semi-evergreen plants only retain some of their leaves for more than one season. See also Deciduous.

Exocarp
The outer layer of the pericarp of a fruit. The exocarp is often thin and hard, or appears to be like a skin. See also Mesocarp, Pericarp.

Family
In plant classification, a group of related genera; the family Rosaceae, for example, includes the genera *Rosa*, *Sorbus*, *Rubus*, *Prunus*, and *Pyracantha*. See also Genus, Species, Subspecies.

Fern
A flowerless, spore-producing plant consisting of roots, stems, and leaflike fronds. See also Frond.

Fiddlehead
The coiled young frond of a fern.

Filament
The stalk that bears the anther in a flower.

Floret
A small flower, usually one of many florets that make up a composite flower such as a daisy.

Flower
The reproductive organ of many plants. It consists of a stem (axis) bearing four reproductive organs: sepals, petals, stamens, and carpels.

Follicle
A dry fruit, similar to a pod, that develops from a single-chambered ovary with one seam that splits open to release the seeds.

Forb
A herbaceous flowering plant that does not have a woody stem and is not a grass, sedge, or rush.

Frond
1. The leaflike organ of a fern. Some ferns produce both barren fronds and fertile fronds, which bear spores. 2. Large, usually compound leaves such as those of palms.

Fruit
The fertilized, ripe ovary of a plant, which contains one or more seeds, such as berries, hips, capsules, or nuts. The term is also used to describe edible fruits.

Fruit set
The process of a flower turning into a fruit after fertilization.

Gametophyte
The sexual reproductive multicellular phase in plants that develops from spores.

Genus (pl. Genera)
A category in plant classification ranked between family and species.

Geophyte
A perennial plant with an underground storage organ, such as a bulb, corm, rhizome, or tuber.

Germination
The physical and chemical changes that take place when a seed starts to grow and develop into a plant.

Gymnosperm
A plant with seeds that develop without an ovary to enclose and protect them while they mature. Most gymnosperms are conifers, whose seeds form on scales and mature within cones.

Habit
The characteristic form of a plant, including its size, shape, and orientation.

Haustorium (pl. Haustoria)
Specialized parasitic plant roots that penetrate the tissues of a host plant and transfer nutrients.

Hemiparasite
A parasite that has green leaves and can photosynthesize, such as mistletoe.

Herbaceous
A non-woody plant in which the upper parts die down to a rootstock at the end of the growing season. The term is mainly used to describe perennial plants, although it also applies to annuals and biennials.

Hermaphrodite
Plant species with flowers in which the male stamens and female pistils are present together in individual bisexual flowers.

Holoparasite
A parasitic plant that cannot photosynthesize and is totally dependent on its host for food and water.

Hybrid
The offspring of genetically different parent plants. Hybrids between species of the same genus are described as interspecific hybrids; those between plants of different, but usually closely related genera are known as intergeneric hybrids.

Inflorescence
A group of flowers borne on a single stem (axis), such as a raceme, panicle, or cyme.

Invasive
A non-native species that is introduced and causes harm to the environment.

Involucre
A ring of leaflike bracts below a flower head.

Labellum
A lip, particularly the prominent third petal of iris or orchid flowers. *See also* Lip.

Lamina
A broad, flat structure; for example, the blade of a leaf.

Leaf
Typically a thin, flat blade (lamina) growing out of a stem that is supported by a network of veins. Its main function is to collect the energy from sunlight that the plant needs in order to photosynthesize. *See also* Leaflet, Photosynthesis.

Leaflet
A subdivision of a compound leaf.

Liana
A woody plant with long, flexible stems rooted in the ground that relies on other plants for support. *See also* Climber.

Lignin
A hard substance in all vascular plants that enables them to stand and grow upright. *See also* Vascular plant.

Lip
A prominent lower lobe on a flower, formed by one or more fused petals or sepals. *See also* Labellum.

Lycophyte
An early form of photosynthetic plant, and the first to evolve with a complex vascular system to transport nutrients and water between the roots and leaves. *See also* Vascular plant.

Maquis
A dense shrubland habitat in the Mediterranean biome, mainly made up of evergreen shrubs and small trees.

Margin
The outer edge of a leaf.

Mesocarp
The middle layer of the pericarp, often the fleshy part of a fruit, such as in a cherry, apricot, or plum. In some pericarps the mesocarp is missing. *See also* Exocarp, Pericarp.

Microphyllous
Describes a plant with microphylls – leaves with a single, unbranched vein.

GLOSSARY

Midrib
The primary, usually central, vein of a leaf.

Monocot (Monocotyledon)
A flowering plant that has only one seed leaf (cotyledon); it is characterized by narrow, parallel-veined leaves. Examples include lilies, irises, and grasses. See also Cotyledon, Eudicot.

Monoecious
A plant with separate male and female flowers that are borne on the same plant. See also Dioecious, Hermaphrodite.

Mycorrhiza
A mutually beneficial (symbiotic) relationship between a fungus and the roots of a plant. See also Symbiosis.

Native
A plant that is indigenous to an area without having been introduced.

Naturalized
A plant that establishes in an area where it is not native, usually after being transported there – intentionally or not – by humans. A naturalized plant becomes part of the local flora and may occasionally become invasive as it escapes the controlling effect of the herbivores, competitors, and diseases of its homeland.

Nectar
A sweet, sugary substance secreted by a nectary, which attracts insects and other pollinators.

Nectary
A gland that secretes nectar to attract pollinators. Nectaries are most frequently located in the flower of a plant, but are sometimes found on leaves or stems. See also Pollination, Pollinator.

Node
A point on a stem from which one or more leaves, shoots, branches, or flowers arise.

Nodule
1. A small knob on a root that contains nitrogen-fixing bacteria. 2. Small swellings on a leaf (on the petiole, midrib, lamina, or margin) that contain bacteria.

Nutrients
Minerals used to develop proteins and other compounds required for plant growth.

Nyctinasty
The movements of plant parts, such as closing petals or moving leaves in response to the cycles of night and day.

Offset
A small plant that develops from a shoot growing out of an axillary bud on the parent plant.

Ovary
The lower part of the carpel of a flower, containing one or more ovules, which may develop into fruit after fertilization. See also Carpel, Pistil.

Ovule
The part of an ovary that develops into the seed after pollination and fertilization.

Palearctic region
A biogeographic region that covers Europe, parts of Asia north of the Himalayas, areas of Africa north of the Sahara, and the northern parts of the Arabian Peninsula.

Panicle
A branched raceme.

Pedicel
The stalk bearing a single flower in an inflorescence.

Perennial
A plant that lives for more than two years.

Perfoliate
Stalkless leaves or bracts that encircle the stem of a plant, so that the stem appears to pass through the leaf blade.

Perianth
The collective term for the calyx and corolla, particularly when they are very similar in form, as in many bulb flowers.

Pericarp
The part of a fruit that develops from the maturing ovary wall. In fleshy fruits the pericarp often has three layers: exocarp, mesocarp, and endocarp. The pericarp of dry fruits is papery or feathery, but on fleshy fruits it is succulent and soft. See also Exocarp, Mesocarp.

Petal
A modified leaf, usually brightly coloured and sometimes scented, that attracts pollinators. A ring of petals on a flower is known as a corolla.

Petiole
The stalk of a leaf.

Phloem
The vascular tissue in plants that conducts sap containing nutrients produced through photosynthesis from the leaves to other parts of a plant.

Photosynthesis
The process by which the energy in sunlight is captured by green plants and used to carry out a chain of chemical reactions to create nutrients from carbon dioxide and water. A by-product of this process is oxygen.

Pinnate
The arrangement of leaflets on opposite sides of the central stalk of a compound leaf.

Pistil
The entire female reproductive organ of a plant. See also Carpel.

Plantlet
A young plant that develops on the leaf of a parent plant.

Pollen
Small grains, formed in the anther of seed-bearing plants, which contain the male reproductive cells of the flower.

Pollination
The transfer of pollen from an anther to the stigma of a flower.

Pollinator
The means by which pollination is carried out; for example, via insects, birds, or wind.

Prickle
A sharp outgrowth from the epidermis or cortex of a plant, which can be detached without tearing the part of the plant from which it is growing.

Propagate
To increase or reproduce plants by seed or by vegetative means.

Propagule
A part of a plant that detaches from the plant to create a new plant. It may be a seed or a clonal piece.

Protocarnivorous
A plant that can trap prey but is unable to digest or absorb it.

Raceme
A cluster of several or many separate flower heads borne singly on short stalks along a central stem, with the youngest flowers at the tip.

Radicle
The root of a plant embryo. The radicle is normally the first organ to appear when a seed germinates.

Refugium (pl. Refugia)
A geographical area where a plant population has been able to survive, or take refuge, from changing conditions.

Relict
A plant that was historically more widespread or diverse and now only exists in a small number of restricted areas.

Resin
A thick, sticky substance formed of organic compounds produced by a tree to heal wounds in its bark that have been inflicted by pests or caused by physical damage.

Respiration
The essential cellular process of breaking down a compound, usually glucose, to release energy to fuel other cellular processes. All living organisms need to respire. *See also* Photosynthesis.

Rhizome
A creeping underground stem that acts as a storage organ and produces shoots at its apex and along its length. *See also* Corm, Tuber.

Rhizosphere
A root system and the substrate that immediately surrounds it.

Root
The part of a plant, normally underground, that anchors it in the soil and through which water and nutrients are absorbed.

Root hair
A threadlike growth that develops behind the root cap. Root hairs extend the surface area of a root and increase the amount of water and nutrients it can absorb.

Rosette
A cluster of leaves radiating from approximately the same point, often at ground level at the base of a very short stem.

Runner
A horizontally spreading, usually slender, stem that runs above ground and roots at the nodes to form new plants. *See also* Stolon.

Sap
The juice of a plant contained in the cells and vascular tissue.

Scale
A reduced leaf, usually membranous, that covers and protects buds, bulbs, and catkins.

Sclerophyllous
Adapted to hot, dry conditions. Sclerophyllous plants typically have tough, often leathery leaves and other adaptations to reduce moisture loss.

Seed
The ripened, fertilized ovule that contains a dormant embryo capable of developing into an adult plant.

Seedling
A young plant that has developed from a seed.

Self-incompatible
A plant that is unable to produce viable seed by fertilizing itself and needs pollen from a different plant in order for fertilization to take place. Also known as "self-sterile".

Self-pollination
The transfer of pollen from the anthers to the stigma of the same flower, or alternatively to another flower on the same plant. *See also* Cross-pollination.

Self-sterile
See Self-incompatible.

Sepal
The outer whorl of the perianth of a flower, usually small and green, but sometimes coloured and petal-like.

Shoot
A developed bud or young stem.

Shrub
A woody perennial plant that has multiple stems. Most shrubs live for several growing seasons.

Spadix
A fleshy flower spike that bears numerous small flowers, usually sheathed by a spathe.

Spathe
A bract that surrounds a single flower or spadix.

Speciation
The evolutionary process by which new, distinct species develop.

Species (sp., pl. spp.)
In plant classification, a group of plants whose members have the same main characteristics and are able to breed with one another, usually sharing a recent common ancestor.

Spike
A flower cluster with individual flowers borne on short stalks or attached directly to the stem.

Spine
A stiff, sharp-tipped, modified leaf or leaf parts such as stipules or petioles.

Sporangium (pl. Sporangia)
A body that produces spores on a plant, such as those on a fern.

Spore
The minute, reproductive structure of flowerless plants, such as ferns, fungi, and mosses.

Spur
1. A hollow projection from a petal or other

GLOSSARY

flower structure, which often produces nectar.
2. A short branch that bears a group of flower buds, such as those found on apple or pear trees.

Stamen
The male reproductive part of a flower comprising the pollen-producing anther and usually its supporting filament or stalk.

Stem
The main axis of a plant, usually above ground, that supports structures such as branches, leaves, flowers, and fruit.

Stigma
The female part of a flower that receives pollen before fertilization. The stigma is situated at the tip of the style.

Stipule
A leafy outgrowth, often one of a pair.

Stolon
A horizontally spreading or arching stem, usually above ground, that roots at its tip to produce a new plant. Often confused with a runner.

Stoma (pl. Stomata)
A microscopic pore in the surface of aerial parts of plants – usually leaves and stems – that allows transpiration to take place.

Style
The stalk that connects the stigma to the ovary.

Submergent
An aquatic plant that lives entirely underwater.

Subspecies (subsp.)
A major division of a species, defining a distinct variant, usually isolated based on geographical location. Subspecies can interbreed successfully with others of the same species.

Subalpine
The region just below the treeline on mountains.

Succulent
A drought-resistant plant with thick, fleshy leaves or stems adapted to store water. All cacti are succulents.

Sucker
A new shoot that develops from the roots or the base of a plant and rises from below ground level.

Symbiosis, symbiotic
Organisms living together in a close, mutually beneficial relationship.

Taproot
The primary, downward-growing root of a plant, such as dandelion.

Tendril
A modified leaf, branch, or stem that is usually long and slender and can attach itself to another plant or object for support.

Tepal
A single segment of a perianth that cannot be distinguished as either a sepal or a petal, as in crocuses or lilies.

Terminal bud
A bud that forms at the apex or tip of a stem.

Testa
The hard, protective coating around a fertilized seed that prevents water from entering the seed until it is ready to germinate.

Thorn
A simple outgrowth from a stem that forms a sharp, pointed end. *See also* Prickle, Spine.

Transpiration
The loss of water by evaporation from plant leaves and stems.

Treeline
The boundary line above which trees are unable to grow due to environmental conditions that are mostly to do with elevation and latitude.

Trichome
Any type of outgrowth from the surface tissue of a plant, such as a hair, scale, or prickle.

Tuber
A swollen, usually underground, organ derived from a stem or a root, that is used to store food to fuel plant growth. *See also* Corm, Rhizome.

Umbel
A flat or round-topped inflorescence in which the flower stalks grow from a single point at the top of a supporting stem. *See also* Inflorescence.

Unisexual
A flower that produces either pollen (male) or ovules (female).

Variegated
Describes irregular arrangements of pigments, usually the result of either mutation or disease, and mainly in leaves.

Vascular plant
A plant that has food-conducting tissues (the phloem) and water-conducting tissues (xylem).

Vicariance
The process by which plant populations become geographically isolated due to environmental factors, such as tectonic movements or climate shifts, and evolve independently, leading to differentiation.

Viviparous
Describes a plant that forms plantlets on leaves, flower heads, or stems. The term may also be used to describe plants that produce bulblets on bulbs.

Whorl
An arrangement of two or more similar organs that all originate from the same zone, whether in circles or spirals.

Winged fruit
A fruit with fine, papery structures that are shaped like wings to help carry the fruits through the air.

Xylem
The woody part of a plant, consisting of supporting and water-conducting vascular tissue.

INDEX

A

Abutilon menziesii 301
Acacia spp. 80–81, 190, 283
Acaciella spp. 80
Acaena alpina 227
Acanthosicyos horridus 197
Acanthus spp. 103, 248
Acer spp. 73, 110
Achillea millefolium 156–157
Aconitum napellus 220
Actinidia chinensis 113
Adansonia spp. 187, 303
Adansonia digitata 90, 166–167
Adenium obesum 180
aerenchyma 247, 254, 291
aerial roots 71, 77, 86, 269
African locust bean 80, 87
African water fern 251
Agathis dammara 78
Agave spp. 172, 173
agricultural areas 58, 192, 274–275, 290–295
  see also crops
Agropyron cristatum 138
Ailanthus altissima 284
akeake 186
Akebia quinata 217
Aldama grandiflora 141
Aldrovanda vesiculosa 304
Alethopteris 29
algae 17, 261, 263, 264
algal blooms 270–271
Allium paradoxum 72
Allocasuarina fraseriana 106
allspice 93
Aloe spp. 174, 175, 200, 201
Aloidendron dichotomum 194
alpenrose 149
alpine bistort 227
alpine meadows 212
alpine mint bush 109
alpine Timothy 145
altitudes 210–213
Aluta maisonneuvei 179
Alysicarpus longifolius 163
amancae 198
Amaranthus caudatus 286

Amazon region 49, 54, 132, 254, 266
Amborella 58
American beech 119
Amorphophallus titanum 303
Andean oak 119
Andira humilis 164
Andromeda polifolia 123
Anemone nemorosa 72
Aneurophyton 26
angiosperms, evolution of 44–47, 51
animals, defences against 141, 158–159
Annonaceae 114
annuals 138
Annularia 31
Antarctica 45, 52, 57
  plants of 142, 145, 146, 264
Antarctic hair grass 145
Anthurium scherzerianum 98
apple 278, 293
aquatic plants 46
  see also wetlands
Aquilegia formosa 220
Araucaria spp. 34, 102, 300
Araucariaceae 40, 43
Arbutus spp. 53, 102
Archaeanthus 44
Archaefructus 46
Archaeopteris 24, 25, 27
Arctic campion 122
Arctic poppies 122
Arctic regions 56, 224, 228
  forests 122, 125, 129
  grasslands 142, 143, 146
Arctium minus 276
argan tree 187
Argyroxiphium sandwicense 213
Arisarum vulgare 102
Aristolochia maxima 89
Armeria maritima 215, 263
Artemisia tridentata 153
Arthraerua leubnitziae 198
Arum maculatum 113
Asarum caudatum 113
ash dieback 100–101
Asimina triloba 114
Asteraceae 140, 231
Astragalus onobrychis 145

Atlantic pearlwort 142
autumn lady's-tresses 154
Avicennia spp. 268–269
Azolla 51
Azorella pedunculata 238

B

Baiera 37
bamboo 164, 217, 291
banana 77, 288–289, 293
Banksia prionotes 190
banyan 71
baobab 90, 166–167, 187, 303
Baragwanathia 20, 22–23
barberries 205, 283
barrels 173
bay laurel 116–117
beach bean 260
beach morning glory 261
beach pea 263
beaked lousewort 142
bear's breeches 103
beech 49, 118–119
Begonia spp. 197, 301
Bennettitales 40, 42
Berberis vulgaris 283
Bertholletia excelsa 54
Berzelia stokoei 153
Betula nana 146
Bignoniaceae 228
big sagebrush 153
Bistorta viviparum 227
Bixa orellana 87
black mangrove 269
black speargrass 163
black wattle 283
bladderwort 254
bloody bellflower 303
bluebell 114, 120
bluebush 187
blue waterlily 250
bog bilberry 126
bog mosses 54
bog plants 246
bog rosemary 123
Bolbitis heudelotii 251
Bombax buonopozense 78

boojum tree 180
Boophone disticha 189
Borassus flabellifer 160
boreal forests 68–69, 122–127
Boswellia sacra 180
Botryopteris 31
Bougainvillea spectabilis 284
bouquet plants 235
Brachychiton rupestris 189
Brachypodium sp. 54
bracts 42
Bridges, Thomas 267
bristlecone pines 176
Briza maxima 139
broadleaves 213
bromeliads 97, 234
Bruguiera cylindrica 248
brush box 92
Buddleja davidii 217, 283
built-up areas 274, 280–285, 299
bulbils 72
bulbs 72, 141
bullhorn acacia 81
bunch grass 237
butcher's broom 106

C

cacao tree 71, 132–133
cacti 86, 177, 206
Cadia purpurea 89
Calamagrostis arenaria 264
Calamites 29, 31
Calamophyton 24, 25, 26
Calamus discolor 75
Calathea sp. 70
Calceolaria spp. 206, 227
Californian laurel 106
Calluna vulgaris 149, 151
Cambrian Period 16
Campanula spp. 129, 206, 219
Canavalia rosea 260
cannonball tree 77
canopy layer 70
Caragana spp. 122, 142
carbon dioxide 18, 192
Carboniferous Period 28–32

INDEX

Cardiocrinum giganteum 219
cardón cactus 206
Carex bigelowii 142
Carmichael, Alexander 156
Carnegiea gigantea 174, 177
carnivorous plants
    forests 87, 98, 125, 126
    wetlands 250, 252, 254
carob 105
carotenoids 73
Castanea sativa 119
castor-oil plant 281
cat claw acacia 81
Catharanthus roseus 70
catkins 118
Cecropia peltata 97
cedar of Lebanon 103
Cedrus spp. 103, 219
Ceiba pentandra 70–71, 84–85, 89
Centaurea nigra 212
century plant 189
Cephalotus follicularis 259
Ceratonia siliqua 105
Ceratophyllum demersum 251
Cereus spp. 172, 173
Ceroxylon quindiuense 97
Chaetolepis microphylla 235
Chamaedendron 25, 26
Chamaenerion angustifolium 125
channelled wrack 261
chaste tree 109
chestnuts 118–119
chickpea 294
Chilean wine palm 105
Chloraea pavonii 197
chlorophyll 50, 70, 73
chocolate 105, 132
cholla 206
Chondrodendron tomentosum 77
Chuquiraga jussieui 235
Cicer arietinum 294
Cinchona pubescens 95
Cistanche tubulosa 179
Cistanthe longiscapa 174, 185
cladodes 185
Cladoxylopsids 24
Clerodendrum trichotomum 108
cliff rose 263
climate change 274–275
    see also Ice Ages
climbing plants, evolution of 31
cloudberry 125
cloud forests 52, 68, 72
    plants of 94–99, 190, 304

cloves 93
clubmosses 22, 26, 28, 30, 36
CO_2 emissions 274–275
coal forests 28–31
coastal deserts 170, 194–199
coastal redwood 105, 301
coastal wetlands 244–245, 260–265
cobra lily 259
coccolithophores 264
Cochemiea poselgeri 175
Cochliosanthus caracalla 90
coconut 294
Cocos nucifera 249, 294
cold deserts 170–171, 202–207
Colicodendron scabridum 190
Colobanthus quitensis 142
Colocasia esculenta 294
Colombian black oak 118
colonies 173
columbines 220
Columnea consanguinea 98
Commiphora myrrha 176
common water crowfoot 254
competition 71, 276–277
cones 34, 39, 47, 73
conifers 253, 263, 301, 303, 304
    evolution of 32–35, 36, 40–41, 45, 52
    in forests 78, 110, 122
conservation 120, 158, 232–233, 300–305
Convallaria majalis 111
Cooksonia spp. 18, 19
Cora glabrata 98
Cornus canadensis 125
Cortaderia selloana 138
cottongrass 56, 58, 146
cotyledons 35
Couroupita guianensis 77
creeping dogwood 125
creosote bush 176
Crescentia cujete 90
Cretaceous Period 44–47, 80, 95
crops 274, 277, 286–287, 290–295
    domestication 58, 278–279
Cuatrecasas, José 230
Cucumis melo 290
cupules 27
cushion plants 212, 215
cuticles 18, 172, 213
cuvie 261

cyanobacteria 23, 271
cycads 36, 40, 42–43, 180, 301
Cyclamen spp. 72, 106
Cynorkis spp. 161, 164
Cyperus papyrus 254, 256–257
Cypripedium calceolus 113
Cytinus hypocistis 106

D

daisies 140, 157, 198, 231
damask rose 294
dandelion 284
Darlingtonia californica 259
Darwin woollybutt 163
Davidia involucrata 114
Dawn redwood 58, 303
Dawsonia superba 95
deciduous leaves 35, 73
decomposers 271
defences 140, 175
Delonix regia 87
Dendrosenecio kilimanjari 213
Deschampsia antarctica 145
desert gold 180
desert heath myrtle 179
desert hyacinth 179
desert ratany 190
desert rose 180
deserts 57, 170–171, 192–193
    adaptations to 172–175
    coastal deserts 194–199
    cold deserts 202–207
    hot dry deserts 176–181
    semi-arid deserts 186–191
Devonian Period 20–27
Devonshire, William Cavendish, Duke of 288
Diapensia lapponica 213
diatoms 263
Dictyophyllum 41
Dictyophyllum nilssonii 42
Dicroidium 36, 37
Didelta carnosa 198
Didierea madagascariensis 187
Digitaria macroblephara 141
Dionaea muscipula 253, 304
Dioonitocarpidium 37
Diospyros lycioides 187
Dipteris conjugata 78
dipterocarps 48, 77, 164
Dipteryx oleifera 304
diseases 100

dispersal 139, 276, 277
Disterigma empetrifolium 238
Dodonaea viscosa 186
dogbanes 190, 198
dollar bush 197
domestication 58, 278–279, 286–287, 290–295
Dorstenia foetida 176
Dracaena cinnabari 179
Dracula spp. 303
dragon blood tree 179
Drepanophycales 22
Drimys granadensis 95
Drosera rotundifolia 126
Druyet Zoubareva, Eugenia 84
Dryas octopetala 224, 276
dry forests 68–69, 86–91
duckweed 46, 253
dune salad bush 198
Dutch elm disease 100
dwarf birch 146
dwarf mobola plum 163
dwarf mountain pine 220
dynamite tree 90

E

early spider orchid 154
East Siberian larch 126
Ebenus cretica 215
Echeveria quitensis 212
Echinopsis candicans 174
Echium wildpretii 213, 220
edelweiss 227
eelgrass 263
einkorn 279, 294
Elaeis guineensis 293
Elkinsia 27
Elphidium crispum 261
Embothrium coccineum 105
emergent layer 70
Emiliania huxleyi 264
Empetrum nigrum subsp. *nigrum* 212
Encephalartos woodii 301
ensets 288
Entada gigas 249, 264
Enterolobium gummiferum 141
Ephedra spp. 53, 205
epiphytes 31, 71, 94
Epipogium aphyllum 125
Equisetum spp. 30, 31, 70, 277
Erica spp. 105, 149

Ericaceae 148–149
Eriophorum spp. 56, 58, 146
Ernestiodendron filiciforme 34
Eryngium spp. 151, 238
Erythrina crista-galli 89
Espeletia spp. 230–231, 238
Eucalyptus spp. 92, 163
Euphorbia spp. 90, 140, 172, 173, 175, 190, 281
Euphrates poplar 205
Euramerica 21, 29
European ash 100
evergreen leaves 73
explosion, dispersal by 277
exposure 212–213
extinction events 36, 38, 48

F

Fabaceae 50–51, 80–81
Fagaceae 49, 118–119
Fagus spp. 52, 73, 119
false wheatgrass 151
Fargesia nitida 217
Federmann, Nikolaus 231
Feijoa sellowiana 93
fernlike plants 24, 25, 26
ferns 78, 111, 237, 251
  evolution of 28, 31, 36–38, 40, 44, 51
  prehistoric 19, 28–29, 33, 36–37, 41, 45
Ferocactus glaucescens 173
fertility 277
fertilizers 271
Ferula assa-foetida 202
Ficaria verna 72
Ficus spp. 55, 70–71, 97, 278, 293
fiddleheads 31
fire, adaptation to 46, 102, 136, 140–141, 154, 161–164
firmosses 238
flame tree 87
floating fern 251
flowering plants, evolution of 17, 27, 44–47, 48, 51
fony baobab 303
forest floor 71
forests 68–69
  adaptations 70–73
  boreal 122–127
  cloud 94–99

evolution of 24–27, 28–31, 48
  temperate 110–115
  tropical and sub-tropical 74–79, 86–91
  see also trees
Fouquieria spp. 173, 180
Fragaria spp. 279
fragmentation 277
frailejónes 230–231, 238
frankincense tree 180
Fraxinus excelsior 100–101
freshwater wetlands 244–245, 246, 250–255
Friar's cowl 102
Fritillaria imperialis 140–141, 150
fronds 31, 179
fruits 54, 80, 249
Fuchsia apetala 94
Fucus sp. 249
fungal diseases 100
fungi 23, 43, 125

G

gametes 19
Gaylussacia brasiliensis 148
Gentiana spp. 225, 237
geophytes 202, 238
Geraea canescens 180
ghaap 198
ghostmen 190
ghost orchid 125
ghost plant 114
giant kelp 264
giant waterlily 266–267
Gigantopteris 33
gin and tonic lobelia 224
ginkgos 283, 298–299
  prehistoric 27, 35, 36, 38–39, 40
Glossopteris 33, 35
Glycine soja 291
golden wattle 81
Gondwana 21, 29, 33, 41, 92, 93
Gossypium hirsutum 294
graceful pine 300
grand mask orchid 153
grape 291
grasslands 136–137, 158, 212
  adaptations to 138–141

cold 142–147
  evolution of 52–55, 57
  temperate 150–155
  tropical and sub-tropical 160–165
grass-of-Parnassus 227
grass pea 293
grazing, defences against 141, 158–159
grendelion 176
Grevillea wickhamii 189
grey mangrove 269
Grimmia longirostris 235
Gronovius, Jan Frederik 129
growth 138–139, 276–277
Guangdedendron 25
guerrilla lycopod 126
gum arabic 80
Gunnera × *cryptica* 284
Guzmania conifera 97
Gymnadenia spp. 146, 224
gymnosperms 44, 53
  see also progymnosperms

H

Halenia weddelliana 238
Haloxylon ammodendron 206
Harmful Algal Blooms (HABs) 271
haustoria 142, 179
heathers 105, 148–149, 151
Heliamphora nutans 258
Helianthus annuus 286
Helleborus foetidus 110
hemiparasites 97, 113, 126, 179
henna tree 179
Heracleum mantegazzianum 277
Herbertus borealis 303
herbs 108–109, 129
Heteropogon contortus 163
Hibiscus tiliaceus 264
Himalayan balsam 220
Himalayan blackberry 284
holly-leaf grevillea 189
holoparasites 205
Hoodia spp. 173, 198
Hooker, William Jackson 267
hop 114
Horneophyton 23
horse mint 109
horsetails 28, 29, 30–31, 40, 41
hot deserts 170–171, 176–181

human environments 274–275, 299
  adaptations 276–279
  agricultural areas 290–295
  protected areas 300–305
  urban areas 280–285
Humboldt, Alexander von 235
Humulus lupulus 114
Huperzia saururus 238
Hura crepitans 90
Huttonaea grandiflora 153
hyacinth 179
Hyacinthoides spp. 72, 114, 120
hybridization 120
Hymenoscyphus fraxineus 100
hyperaccumulators 82
hyphae 23, 43, 125

I

Ice Ages 56–58
Ilex aquifolium 73
Impatiens glandulifera 220, 277
Incarvillea delavayi 228
insulation 213
invasive species 120–121, 280
Ipomoea pes-caprae 261
Iris spp. 145, 246, 253, 304
Ismene amancaes 198
Isoetes spp. 39, 247
isolation 214–215

J

Jacaranda mimosifolia 280
Jamesonia verticalis 237
Japanese evergreen oak 118
Japanese knotweed 111
Japanese maple 110
Jerusalem sage 151
Jubaea chilensis 105
Juglans spp. 49
Jurassic Period 40–43, 78

K

Kalanchoe delagoensis 277
Kalmia latifolia 214
Kamrin, Janice 256

INDEX

kapok tree 71, 84–85
khasi pine 161
Kigelia africana 87
knitted roots 71
Kolpakowsky's tulip 202
Ko'oloa'ula 301
Krameria ixine 190
Krascheninnikovia ceratoides 154

L

Labrador tea 125
Laburnum alpinum 212
Lactuca spp. 279
lady's slipper orchid 113
Lamiaceae 108–109, 140
Laminara spp. 261, 263
Langsdorffia hypogaea 98
Lantana camara 283
larches 35, 126
Larix gmelinii 126
Larrea tridentata 176
latex 74, 190, 202, 293
Lathyrus spp. 263, 293, 304
Laurasia 41
laurel 116
Laurus nobilis 116–117
Lawsonia inermis 179
layers, forest 70–71
leafless milk hedge 190
leaves 70, 73, 172, 213
    evolution of 22, 35
legumes 50, 80, 291, 294
Lemna minor 253, 276
Leontice incerta 205
Leontopodium nivale 227
Lepidodendron 28
lettuce 279
Leucostele atacamensis 206
Lewisia cotyledon 219
Leymus chinensis 151
liana 78, 264
lichens 23, 98, 212
lignotubers 102, 105
lily of the valley 111
Lima orchid 197
lingonberry 143
Linnaea borealis 128–129
Linnaeus, Carl 128, 129
Liriodendron tulipifera 111
little bluestem 154
littleleaf peashrub 142
liverworts 16, 17, 23, 303

Lobelia deckenii 224
Lodoicea maldivica 264
London plane 284
loop-root mangrove 269
Lophostemon confertus 92
Luma apiculata 92
lycophytes 20–23, 24–27, 39, 44, 47
Lycopodium spp. 22, 126

M

Macdonnell Ranges cycad 180
Macrocystis pyrifera 264
Macrozamia macdonnellii 180
Madagascan baobab 187
Madagascan octopus tree 189
Magnolia sprengeri 220
maize 58, 287
Malus domestica 278, 293
mammoth steppe 56, 58
Mangifera indica 293
mango 293
mangroves 49, 264, 268–269
marginal plants 246
marine (coastal) wetlands 244–245, 260–265
Mariosousa spp. 80
marram 264
Meconopsis spp. 146, 214, 227
Mediterranean region 140, 153, 217
    agricultural areas 284, 291, 294
    forests 68–69, 102–107, 116
Medullosa 29
Melampyrum nemorosum 126
Melastomataceae 235
melon 290
Mentha longifolia 109
Mesembryanthemum spp. 172, 174
mesophyll 299
metals 82–83
Metasequoia glyptostroboides 303
methane emissions 275
Metroxylon sagu 78
microphylls 22, 235
mints 108–109
Miocene Period 80
mirror orchid 105
mistletoe 113
Molly the witch 219

monkey puzzle 40, 43, 102, 300
monkshood 220
Monochaetum humboldtianum 235
Monotropa uniflora 114
Monstera deliciosa 77
morning flag 238
moss balls 235
mosses 17, 36, 47, 54, 212
    types 95, 235, 237, 253
mountain houseleek 228
mountains 52, 210–211
    adaptations to 212–215
    high 224–229
    low 216–221
Páramos 234–239
Mucha, Alphonse 116
Musa spp. 77, 293
Musa acuminata 288–289
Muscites 37
mycorrhizae 43, 125
Myoporum sandwicense 161
Myrica gale 246
myrmecochory 106
myrrh tree 176
Myrtaceae 92–93
myrtles 92–93, 179

N

naio 161
namnam bush 179
narcotics 116
needles 32, 73, 213
Nelumbo nucifera 253
Neogene Period 52–55
Nepenthes spp. 98, 258
Nerium oleander 102
Nesiota elliptica 304
Nesocodon mauritianus 303
Neuropteris 28
Nicotiana tabacum 70
Nitraria 49
nitrogen 50, 274
nitrous oxide emissions 275
Noccaea caerulescens 215
Nolana paradoxa 194
non-native species 120, 158, 280
nopal 184
Norfolk Island pine 34
Northern prongwort 303

Nuphar lutea 247
Nymphaea spp. 247, 250
Nymphaeaceae 46–47
Nypa spp. 45, 48, 261

O

oaks 35, 118–119
oil palm 293
oils, defensive 140
Olea europaea 294
Oligocarpia 33
olive 294
Ombalantu baobab 166
Onobrychis cornuta 215
Ophrys spp. 105, 139, 154
opium poppy 293
Opuntia spp. 173, 175
Opuntia ficus-indica 184–185
orchids 197, 224
    in forests 98, 105, 113, 125
    in grasslands 146, 153, 154, 161, 164
Ordovician Period 16–17
Oreopolus glacialis 212
Orthrosanthus chimboracensis 238
Oryza spp. 279, 286, 291
Oshima cherry 114
Osmundaceae 36, 42
ovules 27, 39

P

Pachypodium geayi 190
Pacific madrone 102
Paeonia daurica 219
Paepalanthus spp. 164
Palaquium gutta 74
Paleogene Period 48–51, 54
Palmoxylon 53
palms 48, 75, 78, 97, 105, 293
palmyra palm 160
Pamirian winterfat 154
Pandanus spp. 77, 263
Pangea 29, 32–33, 37, 40–41
papa silvestre 198
Papaver spp. 122, 212, 293
papyrus 254, 256–257
páramo 210–211, 231, 234–239
parasites 179, 190, 205, 302

epiphytes 31, 71, 94
   in forests 98, 106, 113, 114, 125
   hemiparasites 97, 113, 126, 179
parasol pine 106
Parinari capensis 163
Paris polyphylla 216
Parkia spp. 78, 80, 87
Parnassia fimbriata 227
parrot flowers 97
Passiflora foetida 87
Paullinia spp. 78
Paulownia tomentosa 276
pawpaw 114
Paxton, Joseph 266, 288, 289
peas 80–81
peatlands 54, 253
Pedicularis rostrospicata 142
pedunculate oak 119
Pelvetia canaliculata 261
pencil bush 198
Pennisetum villosum 139
permafrost 58, 126
Permian Period 32–35, 36, 39, 42
Pertica 21
   deserts 190, 194, 197, 198
   forests 94, 95, 97
Peruvian bell flowers (*Nolana paradoxa*) 194
Petzke, Karl 132
Philodendron giganteum 71
Phleum alpinum 145
Phlomoides tuberosa 151
Phoenix dactylifera 173
photosynthesis 22, 54, 70, 138–139, 192
phyllode 190
Phyllostachys edulis 291
Phyteuma orbiculare 153
Picea spp. 52, 212
Pilostyles blanchetii 161
Pimenta dioica 93
pines 106, 122, 161, 164, 220, 300
   evolution of 34, 35, 58
Pinguicula vulgaris 251
Pinus spp. 106, 122, 161, 164, 176, 220
Piper auritum 77
pitcher plants 125, 253, 258–259
plantains 288, 293
Platanus spp. 44, 284

Platycerium spp. 78
Pleuromeia 36
Poa spp. 138, 146, 212
Poaceae 52
Podocarpus spp. 45, 263
Podozamites 41
poinsettia 90
poison ivy 111, 293
pollen 27, 39, 89
pollination 139, 174–175, 214, 278
pollution 82, 271, 274, 298–299
Polylepis spp. 227, 237
Polypodium vulgare 111
Polytrichastrum alpinum 212
Populus spp. 114, 205
prairie acacia 80
predators, adaptation to 140–141
prickly pear 184–185
Primula minima 215
Proboscidea louisianica 139
progymnosperms 24, 25, 26
Prosopanche americana 205
Prostanthera cuneata 108–109
Protea repens 154
protected areas 274–275, 300–305
Prunus speciosa 114
Pseudoctenis 41
Pseudotsuga menziesii 73
Psilophyton 21
Psittacanthus cucullaris 97
Pteridium aquilinum 276
Pulsatilla spp. 145, 214
purple saxifrage 228
Puya spp. 172, 234
Pycnandra acuminata 82–83
Pyrenean snowbell 228
Pyrenean violet 217

Q

quaking aspen 114
Quaternary Period 56–58
Queensland bottle tree 189
Quercus spp. 53, 73, 118–119
Quillaja saponaria 106
quillworts 39
Quindio wax palm 97
quiver tree 194

R

Rafflesia arnoldii 303
rainfall 136, 170
rainforests 48, 51, 54, 71
   see tropical biomes, forests
rain tree 89
Ramonda myconi 217
Ranunculus spp. 212, 247, 254
Raoulia eximia 224
rattan 75
red-flowered silk cotton tree 78
red mangrove 264, 269
red oat grass 163
red vanilla orchid 146
redwoods 105, 301
reedmace 251
regeneration 277
resurrection plant 179, 217
Reynoutria japonica 111
rhizobia bacteria 50, 142, 233, 291
rhizoids 20, 23
rhizomes 22, 30–31, 72
Rhizophora spp. 249, 264, 268–269
Rhizophoraceae 49
Rhododendron spp. 125, 149, 217, 284, 304
ribbon roots 98
rice 58, 279, 286, 291
Richea sprengelioides 149
Ricinus communis 281
roots, evolution of 23, 27, 50
root systems 27, 30, 71, 158, 173
   wetlands 247, 249
Rosa × *damascena* 294
Roscoea purpurea 220
round-headed rampion 153
rowan 126
rubber plant 55
Rubus spp. 125, 284
Rudbeckia hirta 153
ruderals 276, 277
Ruscus aculeatus 106

S

sacred lotus 253
sage 108, 151
saguaro cactus 177

Sahagún, Bernardino de 184
sainfoin milk vetch 145
Saint Helena olive 304
saksaul 206
Salicornia europaea 248
Salpiglossis spinescens 206
saltwater wetlands 248–249
   see also coastal wetlands
Salvia spp. 108, 140
Salvinia natans 247, 251
Samanea saman 89
sapote tree 190
Sarracenia purpurea 125, 253, 259
sausage tree 87
Saussurea gossypiphora 228
Sawdonia 20
Saxifraga oppositifolia 228
Scaevola spinescens 180
scales 172
Schivereckia podolica 146
Schizachyrium scoparium 154
Schizanthus hookeri 219
scoggineal 184
screwpine 263
sea cottonwood 264
seasonal adaptation 72–73
sea teak 263
seaweeds 249, 261, 264
seed ferns 28–29, 33, 37
seed plants, evolution of 27, 32, 36–39, 40
seeds 35, 47, 54, 139, 233, 249
Selaginella spp. 47, 179
Selenicereus undatus 86
semi-arid deserts 170–171, 186–191
Sempervivum montanum 215, 228
Senecio cambrensis 283
Senegalia spp. 80, 81, 140
Sequoia spp. 45, 47, 71, 105, 301
sewage 271
Shepherdia canadensis 126
Shorea siamensis 77
shrub layer 71
shrubs 49, 213
Siberian pea tree 122
Sideroxylon spinosum 187
Sigillaria 29
Silene acaulis 122, 212
Silurian Period 18–19, 20, 22
silver hairs 213
silver wattle 190

INDEX

Slavin, Sara 132
slipper flower 206
Smilax aspera 105
snow cover 210
soapbark 106
sohongy 189
soil 82, 138, 215, 233, 246, 299
soil pH 68
Solanum spp. 198, 291
Soldanella villosa 228
Sophora spp. 161, 232–233
Sorbus sudetica 126
Sorghum bicolor 287
soybean 291
Spanish bluebell 120
Spanish moss 95
Sphagnum spp. 54, 237, 253
Sphenopsid 41
spikemosses 47
spines 140, 158, 173, 175
spiny fan-flower 180
Spiranthes spiralis 154
sporangia 18–19
spore plants 17, 28, 36, 47
sporopollenin 17
spring gentian 225
spring pasque flower 145
Spirogyra 17
staghorn ferns 78
Stellaria media 276
stems 18, 70, 172
steppe tundra (mammoth steppe) 56, 58
stiff sedge 142
Stigmaria 30
stink beans 78
stinking passionflower 87
stomata 18, 51, 192, 234
storage organs 72
strangler figs 97
strawberries 279
strobili 22
sub-tropical biomes
  forests 68–69, 74–79, 86–91
  grasslands 160–165
succulent plants 172, 190, 212
sugarbush 154
sundew 126
sunflower 286
sun-pitchers 258
swamp cypress 253
swamps 28–31, 78, 244, 250, 253, 254
sweet chestnut 119

sweet pea 304
symbiosis 23, 43, 50
Syzygium aromaticum 93

T

Taeniophyllum sulawesiense 98
taiga (boreal forests) 68–69, 122–127
tailed snakeroot 113
tall-stilt mangrove 268
tapia tree 164
Tapinanthus oleifolius 179
Taraxacum officinale 284
taro 294
tassel heath 149
Taxodium distichum 253
tea tree 92
Tectona grandis 89
temperate biomes
  forests 68–69, 72, 110–115
  grasslands 136–137, 150–155
temperature, adaptation to 138, 215
teosinte 58
Tephrocactus articulatus 206
Teucrium chamaedrys 109
Themeda spp. 54, 139, 163
Theobroma spp. 70–71, 132–133
thrift 263
tides 244, 249
Tillandsia spp. 71, 95
tillering grass 138
titan arum 303
tonka bean 304
Toxicodendron radicans 111
toxins 215, 271
Tragopogon pratensis 139
Trapa natans 254
travel, dispersal by 276
tree clubmosses 28, 29, 30
tree heather 105
tree of heaven 284
Tree of Life (baobab) 90, 166–167, 187, 303
trees 24–27, 100, 299
  adaptations 70–71, 73, 173, 213, 248–249
Triassic Period 36–39
Trigonobalanus excelsa 118
Trimerophytes 21

Triticum spp. 279, 287, 294
tropical biomes
  forests 68–69, 74–79, 86–91
  grasslands 160–165
tubers 30, 72, 141
Tulipa kolpakowskiana 202
tulip tree 111
tundra 56, 58, 136, 142
tussock grass 138, 146
twinflower 128–129
Typha latifolia 247, 251

U

Uapaca bojeri 164
Ullmannia 33
Umbellularia californica 106
umbrella thorn 189
understorey 70
  deserts 176, 180
  forests 105, 106, 114
upland cotton 294
urban areas 274, 280–285, 299
Utricularia spp. 247, 254

V

Vaccinium spp. 126, 143
Vachellia spp. 80–81, 141, 189
Vanilla planifolia 278
vascular plants 18–19, 23
Venus flytrap 253, 304
Victoria amazonica 254, 266–267
Vietnamosasa pusilla 164
Viscum album 113
Vitex agnus-castus 109
Vitis vinifera 291
Voltzialean conifers 326

W

wall germander 109
walnut 49
water, adaptations to 172–173, 246–249
water caltrop 254
waterlilies 46–47, 250, 266–267

water reservoirs 172
water wheel 304
Weichselia 45
Weinmannia pubescens 97
Welsh ragwort 283
Weltrichia 42
Welwitschia mirabilis 197
western sheoak 106
wetlands 244–245, 271
  adaptations to 215, 246–249
  freshwater 250–255
  marine and coastal 260–265
wheat 58, 279, 287, 294
whisk fern 19
white mangrove 268
Williamsonia 40, 42
willowherb 125
wind pollination 139, 278
Winteraceae 95
Wollemia nobilis 58, 304
woodland 100, 120
Woollsia pungens 148
woolly snowball 228

X, Y, Z

xylem 19, 24
yarrow 156–157
Yaxché (kapok tree) 71, 84–85
Yucca brevifolia 174
Yunnan pine 164
Zea mays 278, 286
Ziziphus nummularia 189
Zostera marina 263
Zosterophyllum 21
Zygnematophyceae 17
Zygophyllum stapffii 197

ACKNOWLEDGMENTS

The publisher would like to thank the following for their kind permission to reproduce their photographs:

(Key: a-above; b-below/bottom; c-centre; f-far; l-left; r-right; t-top)

2-3 Alamy Stock Photo: Florilegius. **4** Michelle EunYoung Song. **5 Alamy Stock Photo:** Penta Springs Limited (br); The Natural History Museum (bc). **6 Alamy Stock Photo:** The Natural History Museum (bl). **Getty Images / iStock:** DigitalVision Vectors / itsme23 (br). **7 Alamy Stock Photo:** The Natural History Museum (bl). **Depositphotos Inc:** mannaggia (br). **8-9 Alamy Stock Photo:** incamerastock / ICP. **11 Alamy Stock Photo:** The Natural History Museum (tl). **Science Photo Library:** Natural History Museum, London (r). **12 Alamy Stock Photo:** Art World. **13 Alamy Stock Photo:** The Natural History Museum. **14-15** Michelle EunYoung Song. **16 Dreamstime.com:** Elif Aytar (cr). **17 Alamy Stock Photo:** PD Archive. **18 MUSE:** Matteo De Stefano (b). **19 Getty Images:** De Agostini via Getty Images / De Agostini (t/x3). **Science Photo Library:** Dr David Furness, Keele University (bl). **22 Alamy Stock Photo:** Album (br). **Dorling Kindersley:** Gary Ombler / Swedish Museum of Natural History (r). **23 Alamy Stock Photo:** Penta Springs Limited / Artokoloro (crb); The Book Worm (tl). **26 Dorling Kindersley:** Gary Ombler / Swedish Museum of Natural History (cra, cr, crb). **Shutterstock.com:** Rawpixel.com (l). **27 Depositphotos Inc:** imagepointfr (cla). **30 Alamy Stock Photo:** The Book Worm. **31 Dorling Kindersley:** Gary Ombler / Swedish Museum of Natural History (tl, tc, tr). **34 Alamy Stock Photo:** Florilegius (b); The Natural History Museum (tr). **35 Alamy Stock Photo:** Jacky Parker (cb); Grethe Ulgjell (clb). **Dreamstime.com:** Patrick Guenette (bc); Bettina Wagner (crb). **Getty Images:** Wild Horizons / Universal Images Group (cla). **38 Dorling Kindersley:** Colin Keates / Natural History Museum, London (cl). **Dreamstime.com:** Evelina Vinogradskaya (r). **Science Photo Library:** Pierre Marchal / Look At Sciences (bl). **39 Alamy Stock Photo:** John Cancalosi (bc); Corbin17 (clb); gameover (cr). **Elsevier:** An Early Triassic Pleuromeia strobilus from Nevada, USA (bl). **42 Dorling Kindersley:** Gary Ombler / Swedish Museum of Natural History (cla, br). **43 Getty Images:** Florilegius / Universal Images Group (l). **Science Photo Library:** Dr Jeremy Burgess (br). **46-47 Dreamstime.com:** Ncl. **46 Shutterstock.com:** Achim Wagner (bl). **47 Alamy Stock Photo:** Penta Springs Limited / Artokoloro (br). **Dorling Kindersley:** Colin Keates / Natural History Museum, London (tr). **Getty Images / iStock:** Hein Nouwens (ca). **50 Alamy Stock Photo:** Art Collection 3 (tl). **50-51 Bridgeman Images:** Photo © The Maas Gallery, London. **51 Alamy Stock Photo:** Sunny Celeste (br). **Dreamstime.com:** Grendachema (tr). **Science Photo Library:** Mark A. Schneider (cr). **54 Alamy Stock Photo:** blickwinkel / AGAMI / W. Leurs (tl); Wild Images (cl); The History Collection (tr/cra). **Dreamstime.com:** Nndanko (b). **Getty Images:** Sepia Times / Universal Images Group (cla). **55 Alamy Stock Photo:** Album. **58 Alamy Stock Photo:** Custom Life Science Images (bc); robertharding / Michael Nolan (tc). **Dreamstime.com:** Karen Black (cb); Evgenii Ivanov (tr). **59 Alamy Stock Photo:** Album. **60-61 Getty Images:** DigitalVision Vectors / bauhaus1000. **64 Alamy Stock Photo:** The Natural History Museum (bl). **Getty Images / iStock:** NSA Digital Archive (br). **Science Photo Library:** Natural History Museum, London (bc). **65 Alamy Stock Photo:** Blueee (tr); The History Collection (tl); Gem Archive (bl); Historic Illustrations (bc); Quagga Media (br). **66-67 Alamy Stock Photo:** The Natural History Museum. **74 Alamy Stock Photo:** Florilegius. **75 Alamy Stock Photo:** Florilegius. **76 Alamy Stock Photo:** Florapix (br); Ch'ien Lee / Minden Pictures (bl). **Dreamstime.com:** Nikolai Kurzenko (cr); Sean Pavone (tl); Leisuretime70 (tr). **Shutterstock.com:** Neenawat Khenyothaa (c). **77 Alamy Stock Photo:** Florilegius (bl). **78 Alamy Stock Photo:** Penta Springs Limited / Artokoloro. **79 Alamy Stock Photo:** blickwinkel / Hauke (tl); Chris Hellier (cr); piemags / nature (bl). **Dreamstime.com:** Lisa Dingo (br). **Shutterstock.com:** Sandeep Gore (bc); Jack Hong (tr). **80 Alamy Stock Photo:** Album (tl). **80-81 123RF.com:** daboost (background texture). **81 Alamy Stock Photo:** Penta Springs Limited / Artokoloro (tr). **82-83** Antony van der Ent, The University of Queensland & Wageningen University: . **84-85 Alamy Stock Photo:** The Natural History Museum. **86 Wikipedia:** creativecommons.org / publicdomain / mark / 1.0. **87 Alamy Stock Photo:** Florilegius (b). **Dreamstime.com:** Chanukakarunarathna4 (c); Elena268 (tc); Kewuwu (cr). **Getty Images:** Moment / Ricardo Lima (tr). **88 Alamy Stock Photo:** Sebastian Kennerknecht / Minden Pictures (tr); Stillman Rogers Photography (br). **Dreamstime.com:** Hellmann1 (b). **Getty Images / iStock:** graphixchon (bl). **Shutterstock.com:** guentermanaus (cr). **89 Wikipedia:** creativecommons.org / publicdomain / mark / 1.0. **90 Alamy Stock Photo:** Album. **91 Alamy Stock Photo:** ffoto_travel (tl). **Getty Images / iStock:** Esin Deniz (b); E+ / guenterguni (tr). **Shutterstock.com:** Ramendri (tc). **92 Alamy Stock Photo:** Album (tr); Penta Springs Limited / Artokoloro (br). **92-93 123RF.com:** daboost (background texture). **93 Alamy Stock Photo:** Album (br); Penta Springs Limited / Artokoloro (tr); Florilegius (bc). **94 Alamy Stock Photo:** Penta Springs Limited / Artokoloro. **95 Alamy Stock Photo:** Album (br); piemags / nature (tl); Allen Creative / Steve Allen (tr); John Richmond (cl). **96 Alamy Stock Photo:** Raquel Mogado (tl); Juergen Ritterbach (bl). **Getty Images / iStock:** cturtletrax (tr). **Shutterstock.com:** guentermanaus (br); Cyrille Redor (cr). **97 Alamy Stock Photo:** Album (br). **98 Alamy Stock Photo:** Album. **99 Alamy Stock Photo:** Florapix (tl); piemags / nature (cl). **Dreamstime.com:** Jessicahyde (tr). **Shutterstock.com:** DadanRamdani94 (tc); Geraldo Morais (b). **100-101 Alamy Stock Photo:** Ashley Cooper pics. **102 Alamy Stock Photo:** Prisma by Dukas Presseagentur GmbH / Heeb Christian (cl). **Dreamstime.com:** Derek Holzapfel (c); ViliamM (cr). **GAP Photos:** Nova Photo Graphik (cra). **103 Alamy Stock Photo:** The Natural History Museum (tr). **Wikipedia:** creativecommons.org / publicdomain / mark / 1.0. **104 Alamy Stock Photo:** Jaime Plaza Van Roon / Auscape (br); Frank Hecker (ca); Fabiano Sodi (br). **Getty Images / iStock:** ffaber53 (tc). **Getty Images:** Stone / Peter Unger (tr); Paroli Galperti / REDA / Universal Images Group (tl). **105 Alamy Stock Photo:** The History Collection (bl). **106 Alamy Stock Photo:** Album. **107 Alamy Stock Photo:** Bob Gibbons (cr). **Dreamstime.com:** Elena Vlasova (b). **Getty Images / iStock:** Gerald Corsi (tr); Margarita Vais (cl); Marina Denisenko (c). **Shutterstock.com:** Edu Mungga (tl). **108-109 123RF.com:** daboost (background texture). **108 Alamy Stock Photo:** Balfore Archive Images (tl); Maidun Collection (tr); Florilegius (br). **109 Alamy Stock Photo:** Album (tr); Penta Springs Limited / Artokoloro (crb); UtCon Collection (bl). **110 Rawpixel:** Medical Botany (1836) by John Stephenson and James Morse Churchill / Rawpixel (l). **Wikipedia:** creativecommons.org / publicdomain / mark / 1.0 (r). **111 Alamy Stock Photo:** David Boag (cr); Nick Kurzenko (tc). **Dreamstime.com:** Orest Lyzhechka (cl); Stanislav Sokolov (tl). **Shutterstock.com:** Harry Thomas Flower (tr). **112 Alamy Stock Photo:** blickwinkel / Cairns (l); Deborah Vernon (tr); Nature Picture Library / Chris Mattison (br). **Dreamstime.com:** Simona Pavan (c). **113 Wikipedia:** (r). **114 Rawpixel:** New York Public Library_Rawpixel. **115 Alamy Stock Photo:** Frank Bienewald (br); SelectPhoto (bl); Sean O'Neill (bc). **Dreamstime.com:** John Hersey (cr); Kristof Lauwers (tr). **GAP Photos:** Andrea Jones (tl). **116 Bridgeman Images:** (bl). **116-117 Alamy Stock Photo:** Album. **118-119 123RF.com:** daboost (background texture). **119 Alamy Stock Photo:** Album (br); Old Images (bl); Penta Springs Limited / Artokoloro (br). **120-121 Fortunato Gatto.** www.fortunatophotography.com. Instagram @fortunato.gatto: . **122 Alamy Stock Photo:** blickwinkel / H. Baesemann (clb). **Dreamstime.com:** Irinav (bc). **Getty Images / iStock:** williamhc (bl). **Wikipedia:** creativecommons.org / licenses / by / 4.0. **123 Alamy Stock Photo:** Penta Springs Limited / Artokoloro. **124 Alamy Stock Photo:** blickwinkel / H. Bellmann / F. Hecker (tr); Christopher Price (cr). **Dreamstime.com:** John1179 (br); Tupungato (tl). **naturepl.com:** Steve Nicholls (bl). **125 Alamy Stock Photo:** Penta Springs Limited / Artokoloro (bl). **126 Wikipedia:** creativecommons.org / licenses / by / 4.0. **127 Alamy Stock Photo:** Brian & Sophia Fuller (bc); Wolfgang Kaehler (tl); Jordana Meilleur (br). **GAP Photos:** Evgeniya Vlasova (cr). **Getty Images:** Photodisc / Ed Reschke (bl). **Shutterstock.com:** Ryzhkov Serhii (tr). **128 Alamy Stock Photo:** Smith Archive (tc). **128-129 Bridgeman Images:** © Purix Verlag Volker Christen. **130-131 123RF.com:** daboost (background texture). **130 Alamy Stock Photo:** Album (br); Florilegius (tl). **131 Alamy Stock Photo:** Heritage Image Partnership Ltd (cra). **132-133 Getty Images:** Moment / PATSTOCK. **134-135 Alamy Stock Photo:** Penta Springs Limited. **142 Alamy Stock Photo:** Colin Harris / era-images (cl); Nature Photographers Ltd / Paul R. Sterry (c). **Dreamstime.com:** Nahhan (cla). **Wikipedia:** creativecommons.org / licenses / by / 4.0 (r). **143 Alamy Stock Photo:** Quagga Media. **144 Alamy Stock Photo:** blickwinkel / Derder (b); Bob Gibbons (tr). **Dreamstime.com:** Viktoria Ivanets (tc); Ihor Martsenyuk (tl). **145 Alamy Stock Photo:** Penta Springs Limited / Artokoloro (r). **146 Wikipedia:** creativecommons.org / licenses / by / 4.0. **147 Alamy Stock Photo:** blickwinkel / R. Bala (tr); piemags / nature (bl). **Dreamstime.com:** Viktor Löki (br). **Getty Images / iStock:** Gerald Corsi (tl). **Shutterstock.com:** Real PIX (tr). **148-149 123RF.com:** daboost (background texture). **Alamy Stock Photo:** Penta Springs Limited / Artokoloro (tc). **148 Alamy Stock Photo:** Florilegius (tl); Penta Springs Limited / Artokoloro (br). **149 Alamy Stock Photo:** History and Art Collection (tr); Pictures Now (crb); Penta Springs Limited / Artokoloro (bc). **150 Alamy Stock Photo:** Penta Springs Limited. **151 Alamy Stock Photo:** Album (br); P Tomlins (ca). **Dreamstime.com:** Tracy Immordino (tl). **iNaturalist.org:** © Urgamal Magsar https://creativecommons.org/licenses/by/4.0 (tr). **152 Alamy Stock Photo:** Linda Freshwaters Arndt (tl); Biosphoto (tr); piemags / nature (bl). **Frank Gaude:** (br). **153 Wikipedia:** creativecommons.org / licenses / by / 4.0 (r). **154 Wikipedia:** creativecommons.org / licenses / by / 4.0. **154 Alamy Stock Photo:** piemags / nature (bl). **Dreamstime.com:** Showface (br); Whiskybottle (tl). **Shutterstock.com:** Danita Delimont (tr). **156-157 Alamy Stock Photo:** Album (c). **Getty Images:** Florilegius / Universal Images Group (t/B). **158-159 Getty Images:** Amy Toensing. **160 Science

• ACKNOWLEDGMENTS •

Photo Library: Natural History Museum, London. **161 Alamy Stock Photo:** Susanne Masters (tr); piemags / nature (tl); Photo Resource Hawaii / Franco Salmoiraghi (tc). **iNaturalist.org:** Nico Blüthgen https://creativecommons.org/publicdomain/zero/1.0 (cl). **Wikipedia:** creativecommons.org / licenses / by / 4.0 (br). **162 Alamy Stock Photo:** Papilio / Michael Maconachie (b). **Shutterstock.com:** Hank Asia (tc); Mang Kelin (tl). **Nicholas Case Wightman:** (tr). **163 Wikipedia:** creativecommons.org / licenses / by / 4.0. **164 Alamy Stock Photo:** Album. **165 Alamy Stock Photo:** piemags / nature (cra, b). **Zoë Goodwin:** (tr). **Shutterstock.com:** AlivePhoto (tl). **Wikipedia:** João de Deus Medeiros https://creativecommons.org/licenses/by/2.0 (tc). **166-167 Rawpixel. 168-169 Alamy Stock Photo:** The Natural History Museum. **176 Alamy Stock Photo:** Adam Jones / DanitaDelimont (c); piemags / nature (cl); Nature Picture Library / Chris Mattison (clb); Florilegius (bl). **177 Alamy Stock Photo:** Historical image collection by Bildagentur-online. **178 Alamy Stock Photo:** Iain Lowson (br); Tessa Pietersma (cl). **Bridgeman Images:** Photo © Michel Viard / Horizon Features. All rights reserved 2025 (bl). **Shutterstock.com:** SC Gardens (tl); Pawel Uchorczak (tr). **179 Alamy Stock Photo:** Florilegius (br). **180 Alamy Stock Photo:** Album. **181 Alamy Stock Photo:** David Massemin / Biosphoto (tr); piemags / nature (c); Kevin Schafer (bl). **Dreamstime.com:** Vladimir Melnik (tl); Oksanaphoto (br). **182-183 123RF.com:** daboost (background texture). **Alamy Stock Photo:** Album (tc). **182 Alamy Stock Photo:** Art Collection 3 (tl). **183 Alamy Stock Photo:** Penta Springs Limited / Artokoloro (c). **184 Getty Images:** Moment / © fitopardo (cra). **184-185 Bridgeman Images:** The Stapleton Collection. **186 Alamy Stock Photo:** Heritage Images. **187 Alamy Stock Photo:** piemags / nature (br). **Dreamstime.com:** Gerold Grotelueschen (cr). **Getty Images / iStock:** Leonid Andronov (bl). **188 Alamy Stock Photo:** Stephanie Jackson - Australian landscapes (bl); Suzanne Long (tl); Chris Hellier (tr); Ariadne Van Zandbergen (br). **Dreamstime.com:** Arif Rahman Awahab (cl). **189 Bridgeman Images:** © Florilegius. **190 Alamy Stock Photo:** Album (bl). **191 Alamy Stock Photo:** Stephanie Jackson - Aust wildflower collection (tc); Nature Picture Library / Lorraine Bennery (tl); Gabbro (br). **Dreamstime.com:** Walter Medina (bl). **Shutterstock.com:** EQRoy (tr). **192-193 UN/DPI Photo:** Marco Domino. **194 Wikipedia:** creativecommons.org / licenses / by / 4.0. **195 Alamy Stock Photo:** Florilegius. **196 Alamy Stock Photo:** Michael & Patricia Fogden / Minden Pictures (tl). **Getty Images / iStock:** Gerald Corsi (bl). **iNaturalist.org:** Paúl Gonzáles https://creativecommons.org/licenses/by/4.0 (r). **Science Photo Library:** Francesco Tomasinelli (cl). **197 Alamy Stock Photo:** Florilegius (b). **198 Artvee:** Priscilla Susan Bury (1799-1872). **199 Alamy Stock Photo:** Ann Miles (t); piemags / nature (bc). **Dreamstime.com:** Grobler Du Preez (br). **iNaturalist.org:** Antonio W. Salas https: / / creativecommons.org / licenses / by / 4.0 (bl). **200-201 123RF.com:** daboost (background texture). **200 Alamy Stock Photo:** The Picture Art Collection (tr). **201 Alamy Stock Photo:** Penta Springs Limited / Artokoloro (bc). **202 Alamy Stock Photo:** Florilegius. **203 Bridgeman Images:** © Purix Verlag Volker Christen. **204 Dreamstime.com:** Niuniu (b). **Getty Images / iStock:** Max Zolotukhin (tl). **Lytton John Musselman:** (cra). **Shutterstock.com:** photowind (tr). **206 Dreamstime.com:** Sl Photography. **207 Alamy Stock Photo:** Gina Kelly (br). **iNaturalist.org:** © Nolan Exe https://creativecommons.org/licenses/by/4.0 (bl); © Álvaro Parra Valdivia https://creativecommons.org/licenses/by/4.0 (cr). **Shutterstock.com:** Frank Wagner (t). **208-209 Getty Images / iStock:** DigitalVision Vectors / itsme23. **216 Wikipedia:** creativecommons.org / licenses / by / 4.0. **217 Dreamstime.com:** Jamesjrrassaerts (ca); Neydtstock (tc); Wirestock (tr); Matauw (cra). **Wikipedia:** creativecommons.org / licenses / by / 4.0 (b). **218 Dreamstime.com:** Lostafichuk (tl); Christian Weiß (tr); Satish Parashar (br). **Getty Images / iStock:** dentdelion (cr); Dr John A Horsfall (bl). **219 Alamy Stock Photo:** History and Art Collection (r). **220 Alamy Stock Photo:** The Print Collector. **221 Alamy Stock Photo:** Anne Gilbert (tl); Michael Wheatley (tr); Nature Picture Library / Felis Images (bl); Nature Picture Library / Nick Upton (bc). **Dreamstime.com:** Rob Lumen Captum (br). **Shutterstock.com:** lorenza62 (cl). **222-223 123RF.com:** daboost (background texture). **222 Alamy Stock Photo:** Historic Collection (tr); Penta Springs Limited / Artokoloro (tl). **Getty Images:** Moment / mikroman6 (br). **223 Alamy Stock Photo:** Library Book Collection (cra); Penta Springs Limited / Artokoloro (bl); PhotoStock-Israel / Botanical Illustration (crb). **224 Alamy Stock Photo:** Frédéric Grimaître (cb). **Dreamstime.com:** Soloway (br). **Getty Images:** Auscape / Universal Images Group (bc). **Wikipedia:** creativecommons.org / licenses / by / 4.0. **225 Getty Images / iStock:** DigitalVision Vectors / itsme23. **226 Alamy Stock Photo:** Arterra Picture Library / Clement Philippe (cl); Rudi Sebastian (tl); Linda Reinink-Smith (bl); blickwinkel / S. Derder (bc); Florapix (br). **iNaturalist.org:** Matt Berger https://creativecommons.org/licenses/by/4.0 (tr). **227 Wikipedia:** creativecommons.org / licenses / by / 4.0. **228 Wikipedia:** creativecommons.org / licenses / by / 4.0 (l). **229 Alamy Stock Photo:** blickwinkel / McPHOTO / BRS (br); imageBROKER / dad fotos (tc); Fabiano Sodi (bl). **Dreamstime.com:** Stefans42 (tr). **Shutterstock.com:** Jiang Tianmu (tl). **230-231 Alamy Stock Photo:** Album. **232-233 Toromiro, Sophora toromiro (Phil.) Skottsb., collected 31 March 2022, cultivated Wellington., New Zealand. CC BY 4.0. Te Papa (SP114679):** . **234 Alamy Stock Photo:** Penta Springs Limited / Artokoloro (r). **235 Alamy Stock Photo:** piemags / nature (tr, cra). **Dreamstime.com:** S Billingham (tc). **Wikipedia:** creativecommons.org / licenses / by / 4.0 (b). **236 Alamy Stock Photo:** Danita Delimont (tc). **Dreamstime.com:** Henri Leduc (b). **Getty Images / iStock:** Gerald Corsi (tl); Antonio Duarte (tr). **237 Wikipedia:** creativecommons.org / licenses / by / 4.0. **238 Wikipedia:** creativecommons.org / licenses / by / 4.0. **239 Alamy Stock Photo:** piemags / nature (cr). **Dreamstime.com:** Iryna Kurilovych (bl); Jinfeng Zhang (tr). **naturepl.com:** Pete Oxford (br). **Science Photo Library:** Sinclair Stammers (tl). **Wikipedia:** LUFJARD/https://creativecommons.org/licenses/by/4.0/deed.en (bc). **240-241 123RF.com:** daboost (background texture). **240 Alamy Stock Photo:** Penta Springs Limited / Artokoloro (tl); The Picture Art Collection (br). **241 Alamy Stock Photo:** Florilegius (tr); Penta Springs Limited / Artokoloro (tl). **Getty Images / iStock:** DigitalVision Vectors / ZU_09 (br). **242-243 Alamy Stock Photo:** The Natural History Museum. **250 Wikipedia:** creativecommons.org / publicdomain / zero / 1.0. **251 Alamy Stock Photo:** Alain Kubacsi / Biosphoto (tr); Alessandro Mancini (c); blickwinkel / Teigler (cr). **Dreamstime.com:** Oleh Marchak (tc). **Wikipedia:** creativecommons.org / licenses / by / 4.0 (b). **252 Alamy Stock Photo:** Mark Conlin (tl); Anne Gilbert (tr). **Dreamstime.com:** Simona Pavan (bc); Wentong Wang (cl); Taina Sohlman (bl). **Getty Images / iStock:** ArendTrent (br). **253 Getty Images / iStock:** NSA Digital Archive. **254 Alamy Stock Photo:** imageBROKER.com / BAO (tr). **255 Alamy Stock Photo:** Arterra Picture Library / De Meester Johan (tl); Angus McComiskey (tr); Manfred Ruckszio (cl); Paulette Sinclair (b). **256-257 Yale Center for British Art, Paul Mellon Collection. 258-259 123RF.com:** daboost (background texture). **Alamy Stock Photo:** Quagga Media (bc). **258 Alamy Stock Photo:** The History Collection (tl); The Natural History Museum (tr). **259 Alamy Stock Photo:** Florilegius (bc); The Natural History Museum (br). **Shutterstock.com:** Rawpixel.com (tr). **260 Alamy Stock Photo:** The Natural History Museum. **261 Alamy Stock Photo:** Florilegius (tr); imageBROKER.com / Horst Mahr (bl). **Dreamstime.com:** Evgeniy Fesenko (br). **Getty Images:** Alexis Rosenfeld (cb). **Science Photo Library:** Power And Syred (clb). **262 Alamy Stock Photo:** Jake Davies (bc); Alf Jacob Nilsen (tr); Nature Photographers Ltd / Paul R. Sterry (cl). **Dreamstime.com:** Whiskybottle (tl). **Shutterstock.com:** Chris Ison (br); NPvancheng55 (bl). **263 Alamy Stock Photo:** Science History Images. **264 Wikipedia:** creativecommons.org / licenses / by / 4.0 (r). **265 Alamy Stock Photo:** Nature Picture Library / Martin Gabriel (bl); Steve Gschmeissner / Science Photo Library (br). **Dreamstime.com:** Ethan Daniels (tc); Kagab4 (tl); Delstudio (tr, cr). **266-267 Wikipedia:** creativecommons.org / licenses / by / 4.0. **266 Bridgeman Images:** The Board of Trustees of the Royal Botanic Gardens, Kew (cl). **268 Alamy Stock Photo:** Book_Worm (br). **268-269 123RF.com:** daboost (background texture). **Alamy Stock Photo:** Historic Collection (tc). **269 Alamy Stock Photo:** Archivah (br); Historic Collection (tr). **Shutterstock.com:** Rawpixel.com (bl). **270-271 Daniel Núñez:** . **272-273 Depositphotos Inc:** mannaggia. **280 Alamy Stock Photo:** Les Archives Digitales. **281 Alamy Stock Photo:** Sunny Celeste. **282 Alamy Stock Photo:** Botany vision (tr); Neil Maclachlan (tl). **Dreamstime.com:** Meunierd (br). **Floral Images:** Floral Images (bc). **Shutterstock.com:** Dreey Photography (bl). **283 Alamy Stock Photo:** Florilegius. **284 Alamy Stock Photo:** Historic Images (t). **285 Alamy Stock Photo:** blickwinkel / Jagel (bl); Nature Photographers Ltd / Paul R. Sterry (tl); Nik Taylor (tc). **Dreamstime.com:** Simona Pavan (tr). **Getty Images:** Bluered / REDA / Universal Images Group (br). **Getty Images / iStock:** Alexandr Yakovlev (cl). **286 Alamy Stock Photo:** The Natural History Museum (tl, br). **286-287 Alamy Stock Photo:** Penta Springs Limited / Artokoloro (tc). **287 Alamy Stock Photo:** Axis Images (tr); Skimage (bl); Danvis Collection (br). **288 Alamy Stock Photo:** The Picture Art Collection (bl). **288-289 Bridgeman Images:** © Fitzwilliam Museum. **290 Alamy Stock Photo:** Album (bl). **291 Alamy Stock Photo:** Nigel Cattlin (tl). **Dreamstime.com:** Hel080808 (tr). **Getty Images / iStock:** alika1712 (cl). **Science Photo Library:** Dr Jeremy Burgess (cr). **Shutterstock.com:** Gaston Cerliani (tc). **292 Alamy Stock Photo:** blickwinkel / McPHOTO / HRM (cl); Chris Mattison (tr); Alan Skyrme (bl); Susanne Masters (bc). **Dreamstime.com:** Aliaksandr Mazurkevich (br). **Getty Images / iStock:** mgfoto (tl). **293 Alamy Stock Photo:** Old Images. **294 Alamy Stock Photo:** Old Images. **295 Alamy Stock Photo:** blickwinkel / R. Koenig (cl); Joerg Boethling (tl). **Depositphotos Inc:** jukree (br). **Dreamstime.com:** Sahil Ghosh (crb); Horst Lieber (tr). **Getty Images / iStock:** E+ / cinoby (bl). **296 Alamy Stock Photo:** Florilegius (br); Well / BOT (tl). **296-297 123RF.com:** daboost (background texture). **Getty Images:** Hulton Archive / Ann Ronan Pictures / Print Collector (tc). **297 Alamy Stock Photo:** Quagga Media (bl); The Natural History Museum (tr). **Getty Images:** Florilegius / Universal Images Group (br). **298-299 Getty Images:** Visual China Group / VCG. **300 Wikipedia:** creativecommons.org / licenses / by / 4.0. **301 Alamy Stock Photo:** John Bracegirdle (cla); David Cobb (tc); World History Archive (bl). **Getty Images / iStock:** Samuel Howell (tl). **302 Alamy Stock Photo:** Cyril Ruoso / Minden Pictures (tr); Top Photo Corporation (tl); Survivalphotos (bl); Zoonar / Peter Himmelhuber (bc). **Sharon Pilkington:** British and Irish Bryological Society (cl). **Shutterstock.com:** Bpk Maizal (br). **303 Wikipedia:** creativecommons.org / licenses / by / 4.0. **304 Wikipedia:** creativecommons.org / licenses / by / 4.0. **305 Alamy Stock Photo:** Wiskerke (tr). **Getty Images:** Stone / Paul Starosta (br). **Evan Gora:** (bl). **Jim Murrian:** (tl). **Robin Whiting:** RHS Rhododendron, Camelia & Magnolia Group (tc)

Cover images: Front: **Alamy Stock Photo:** Album c, The Natural History Museum tr; **Bridgeman Images:** br; Back: **Alamy Stock Photo:** Album bl, The Natural History Museum r; Spine: **Alamy Stock Photo:** The Natural History Museum

DK LONDON

Senior Art Editor Duncan Turner
Senior Editor Helen Fewster
Editors Annie Moss, Tom Booth, Nathan Joyce
Designer Judy Caley
Managing Editor Angeles Gavira
Managing Art Editor Michael Duffy
Production Editor Robert Dunn
Senior Production Controller Meskerem Berhane
Publishing Director Georgina Dee
Art Director Maxine Pedliham

DK DELHI

Senior Editor Anita Kakar
Senior Jacket Designer Suhita Dharamjit
Senior DTP Designer Harish Aggarwal
Senior Jackets Coordinator Priyanka Sharma Saddi

First published in Great Britain in 2026 by
Dorling Kindersley Limited
20 Vauxhall Bridge Road,
London SW1V 2SA

The authorized representative in the EEA is
Dorling Kindersley Verlag GmbH. Arnulfstr. 124,
80636 Munich, Germany

Copyright © 2026 Dorling Kindersley Limited
A Penguin Random House Company
10 9 8 7 6 5 4 3 2 1
001–350181–May/2026

All rights reserved.
No part of this publication may be reproduced, stored in or introduced into a retrieval system, or transmitted, in any form, or by any means (electronic, mechanical, photocopying, recording, or otherwise), without the prior written permission of the copyright owner.

DK values and supports copyright. Thank you for respecting intellectual property laws by not reproducing, scanning or distributing any part of this publication by any means without permission. By purchasing an authorized edition, you are supporting writers and artists and enabling DK to continue to publish books that inform and inspire readers.

No part of this publication may be used or reproduced in any manner for the purpose of training artificial intelligence technologies or systems. In accordance with Article 4(3) of the DSM Directive 2019/790, DK expressly reserves this work from the text and data mining exception.

A CIP catalogue record for this book
is available from the British Library.
ISBN: 978-0-2417-4509-0

Printed and bound in China

www.dk.com

ACKNOWLEDGMENTS

DK would like to thank illustrators Andrew Beckett, Dan Crisp, and Laura Silburn, picture researcher Jackie Swanson, and cartographer Mike Hall. DK would also like to thank Adam Brackenbury for image retouching, Joy Evatt for proofreading, Elizabeth Wise for creating the index, and fact-checkers Bharti Bedi, Karen Font, Kris French, Michelle Harris, Priyanka Kharbanda, Angela Modany, and Avery Naughton. Design assistance by Francis Wong and Izzy Poulson, editorial support by Anna Cheifetz, Jane Simmonds, Miezan Van Zyl, and Hannah Westlake, and pre-production support by Tarun Sharma and Umesh Singh Rawat.

AUTHORS

Dr Gregory J. Kenicer, Lead author, is a botanist and lecturer at the Royal Botanic Garden Edinburgh. His research interests include the fascinating evolution and diversity of peas and beans, and the profound connections between plants and humanity.

Dr Zoë Goodwin is the Postgraduate Co-ordinator for the Royal Botanic Garden Edinburgh's MSc Biodiversity and Taxonomy of Plants programme. She specializes in the identification of tropical plant families.

Dr Alexander J. Hetherington is an evolutionary palaeobiologist who holds positions as a Future Leaders Fellow and Senior Lecturer at the University of Edinburgh. He is also a Research Associate at the Royal Botanic Garden Edinburgh and the National Museums Scotland.

Dr Jess Rickenback is a Lecturer in Physical Geography at the University of Edinburgh and a Research Associate at the Royal Botanic Garden Edinburgh. Her work focuses on plant biogeography.

Dr Louis Ronse De Craene is a botanist and Research Associate at the Royal Botanic Garden Edinburgh. His research interests include floral morphology, and the evolution of flowers.

Dr Michael Sundue is an Integrative Taxonomist at the Royal Botanic Garden Edinburgh. He specializes in the systematics, evolution, and ecology of ferns and lycophytes.

The Royal Botanic Garden Edinburgh is a leading botanic garden and a global centre for biodiversity science, horticulture and education. The organization conserves one of the world's richest botanical collections across four sites in Scotland – at Edinburgh, Benmore, Dawyck, and Logan. Through cutting-edge collaborative science, conservation efforts, and educational programmes the Royal Botanic Garden Edinburgh is working to create a positive future for plants, people, and the planet.

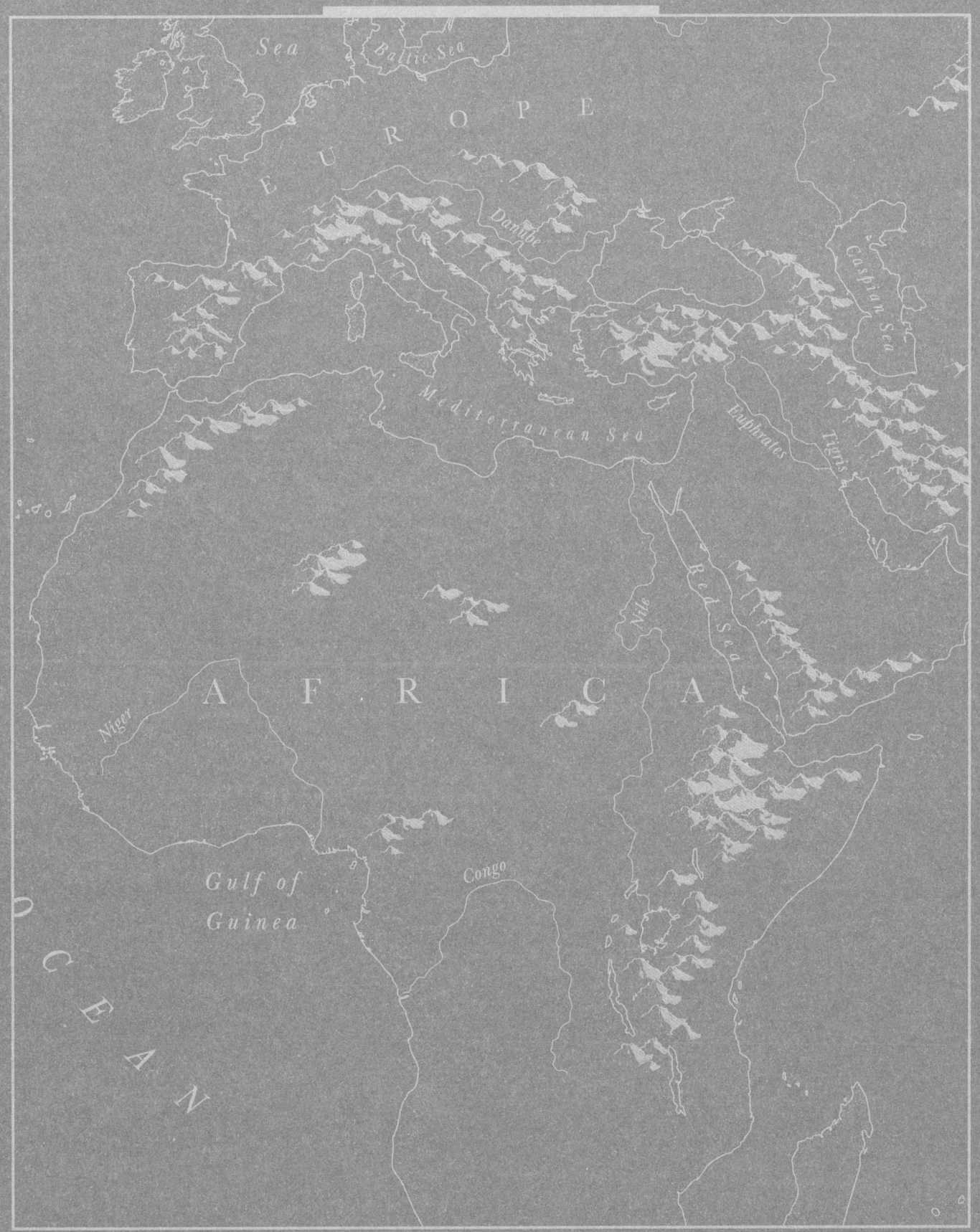

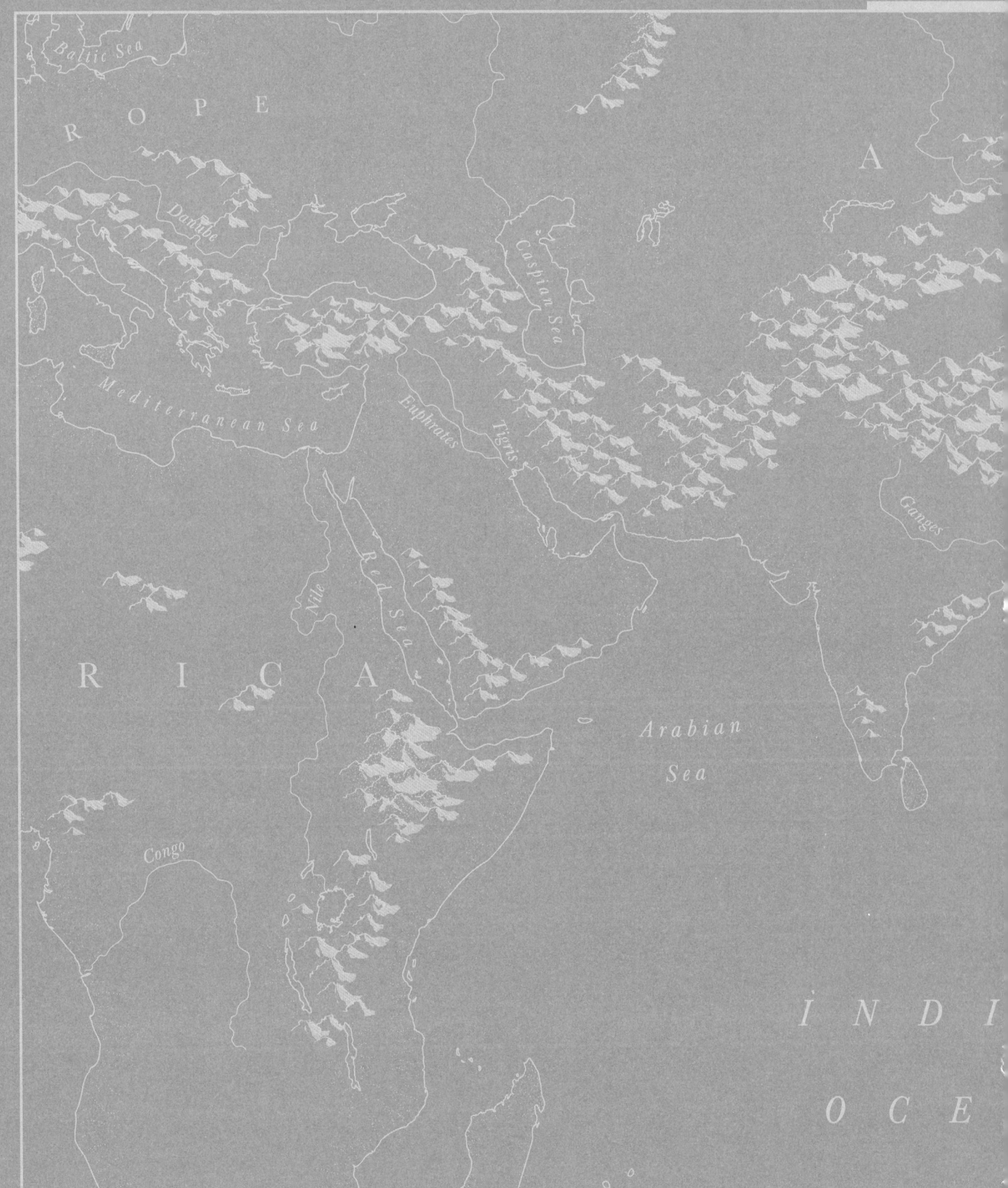